THE SOVIET SASQUATCH

Boris Porshnev

Translated by Lars Thomas

Edited by Chris Clark

Typeset by Jonathan Downes,
Cover and Layout by SPiderKaT for CFZ Communications
Using Microsoft Word 2000, Microsoft Publisher 2000, Adobe Photoshop CS.

First published in Great Britain by CFZ Press

CFZ Press
Myrtle Cottage
Woolsery
Bideford
North Devon
EX39 5QR

ISBN: 978-1-909488-64-9

Translator's Preface

Translating this book has not been an easy task. Partly because the author's style is not exactly what one would call light reading, and partly because the book was very clearly written at a time when it must have been extremely important to observe every detail of the political protocol. So for the sake of readability, I have been forced to do a certain amount of editing. I have tried very much to preserve the spirit of the original, but in many cases I have been forced to straighten the author's long and convoluted sentences to give them some resemblance to modern English.

As this book was originally written at a time when the Soviet Union was still in existence, there are lots of references to former Soviet areas. I have made no attempt to change these according to contemporary political boundaries as this would change the style of the book too much.

It is quite clear that this manuscript had not been completely finished by the author. There are various notes spread throughout the text about having to check footnotes and correct errors. This among other things means it is possible there are factual errors here and there in the text. There are also rather strange jumps in the narrative, which I can only surmise is due to parts of the text that are actually missing. Unfortunately, there is nothing much I can do about that, but it is my distinct impression that the amount of errors is actually minimal.

One major flaw in the unfinished state of the manuscript is the lack of an index. I can only suppose the author never got around to making one.

To keep the number of linguistic errors to a minimum, I have consulted a number of students at the Department of Cross-Cultural and Regional Studies at the University of Copenhagen, and a couple of Russian-speaking friends. I am deeply grateful for their time and effort, but I must stress that any errors still lurking about the pages are entirely my fault.

Lars Thomas, Copenhagen 2017

Editor's Preface

Porshnev uses the terms Bigfoot, yeti, snowman, relict hominid etc. more or less indiscriminately throughout the text, consistent with his view that ultimately they all refer to the same or closely related creatures. The word hominid here means a creature in the form of or looking like a human, rather than the technical sense used in zoology.

As noted above, the book had no index, and I have attempted to supply one.

Porshnev either had access to poor translations of his English-language sources or was quoting from memory with the intention of revising later; when I came to compare his text with the original versions the discrepancies are sometimes enough to change the sense altogether. I have resisted any temptation to restore the originals, merely adding corrections in the notes.

I have made no attempt to impose consistency on place names. For one village in Tibet, the Internet suggested twelve variant spellings, and Google Maps a thirteenth! Such well-known places as Rhodesia, Leningrad and Peking retain the names they had when Porshnev wrote. For every identifiable location I have chosen a form that can be found in such standard reference sources as Wikipedia and Google Maps. The maps identify the major areas of research, and the index gives clues to the location of some unfamiliar places. Of course, there remain a number of villages and geographical features that are impossible to identify. Some places, such as Altyn Mazar near the spot on the Fedchenko Glacier where Pronin had his famous sighting, no longer exist.

Writing for a small academic audience, Porshnev added a great number of footnotes throughout the text: 400 in the first eight chapters alone. This was never completed – many of them are notes to himself to add further data, while others are empty numbers where the author never got around to finding the exact references – and the great majority are to Russian sources, in particular the Information Materials of the Commission, which are now wholly inaccessible. Only a handful have been retained. All references to books are to the English versions and not the translations Porshnev used.

The names of some people who might have been household names to Porshnev's Russian audience are nearly unknown in the West. In those cases I have added a few details in the notes.

Chris Clark, 2018

Foreword

For many years, I have indulged my twin passions for cryptozoology and (what I call) rock and roll archaeology. As such, I expended lots of time, effort and money in tracking down obscure records by artists whose work I admire. Often, these records were more or less illegal, having been purloined from the cutting room floor of various recording studios across the world. These ‘bootleg’ recordings often included valuable insights into the composition and recording process, and it has gratified me that in recent years, various artists such as Bob Dylan, David Bowie, and The Beatles, have issued these historic musical documents officially.

One of my oldest friends is a bloke called Rob Ayling, who runs a record company that – over the years – has released some magnificently obscure artefacts. In many ways, I based my business model upon his when I set up CFZ Press, where as well as publishing cutting edge research in the cryptozoology and allied disciplines, I wanted to track down some of the legendary books that have been written on these subjects but which, for various reasons, either remained unpublished or regrettably obscure. Let me introduce you to Boris Porshnev.

Boris Fyodorovich Porshnev (Russian: Бори́с Фёдорович По́ршнев; 7 March [O.S. 22 February] 1905, Saint Petersburg – 26 November 1972, Moscow) was a Soviet historian known for his works on popular revolts in Ancien Régime France and a doctor of social sciences working on psychology, prehistory, and neurolinguistics as relating to the origins of man. (From Wikipedia, the free encyclopedia).

Within cryptozoology, Porshnev is best known for having co-authored *L'homme de Néanderthal est toujours vivant* (1974). He also was rumoured to have written an unpublished book about BHM phenomena in the Soviet Union. And, it impressed me greatly when Richard Freeman managed to get hold of a copy of the unpublished manuscript (written in Russian), from Porshnev’s family. And even more impressively, he got permission for us to publish it. This was several years ago, but we passed it over to Danish polymath, Lars Thomas, who - amongst lots of other things - can read Russian and he painstakingly translated it, before passing it over to Dr Chris Clark for editing.

Lars writes: “I just wanted to emphasize yet again - that this is not a finished book. It is/was a work in progress when Porshnev died, and I think it is very important, that you tell people that, when you start making press releases and so forth. There are bits missing in the manuscript, things he never got around to write, and things that are only notes and sketches, some of which I have omitted completely, because they have no value. We can't have people expecting a major finished work in the BHM-field, because it is in fact a historically interesting document with a lot of flaws, but I suspect also with a lot of hitherto unknown information.”

I would like to thank everybody for their immensely hard work on this project and I am very proud that the CFZ has managed to bring you this peerless piece of cryptozoological history.

Enjoy.

Jon Downes
(Director, CFZ)

Boris Fyodorovich Porshnev
(Russian: Бори́с Фёдорович По́ршнев)
(1905 - 1972)

Contents

Acknowledgements

This book could not have been written without the assistance of a number of people. The prominent Soviet zoologists: L.P. Astanin, N.I. Burchak-Abramovich, G.P. Dementiev, N. Ladygina-Kotc, A.A. Mashkovtsev, M.F. Nesturkh, P.P. Smolin, V.A. Khakhlov, ethnographer S. Tokarev and paleontologist V.I. Gromov all gave valuable advice or contributed material. The author would like to express his gratitude towards his colleagues, but would also like to make it clear that he alone is responsible for the text.

The author gratefully acknowledges the assistance of the following foreign colleagues: Bernard Heuvelmans (Belgium), Ivan Sanderson (Scotland, USA), George Agogino (USA), Peter Byrne (USA), Corrado Gini (Italy), Hans Pettssha (German Democratic Republic), Rinchen (Mongolian People's Republic).
My long-standing employees A.A. Shmakov, V.A. Telisheva, J.J. Kofman, V.L. Bianchi, and T.N. Dunaevskaya kindly read the manuscript and shared their observations.

This research would not have been possible without a huge number of correspondents and their sightings.

Readers who do not have the time or the inclination to familiarise themselves with the lengthy descriptive material can limit their reading to Chapters 1, 5, 11, 12, 13, and the final discussion.

The book reflects my knowledge and understanding of the subject by January 1963. When it is published it will probably already be out of date, as the accumulation and analysis of data is a continuing process.

B. Porshnev
Moscow, January 1963

Introduction

Editor's note. *Porshnev begins his book with a section so filled with technical terms and obscure words that the reader, imagining the rest of the book to be on the same level, may be reluctant to go further. A few definitions have been added here to avoid the need for continuous reference to Wikipedia by the reader. It is worth stressing that views on Neanderthals and their place in the evolutionary scheme have changed considerably since Porshnev wrote.*

Acheulean: refers to a characteristic set of stone tools rather than a particular time or place. Produced in the Lower Paleolithic and associated with *Homo erectus* fossils, they appear about 1,750,000 BP and are very widely distributed.

Anthropogenic: relates here to those sciences relevant to the study of human origins, not to those concerning human impact on the environment. Presumably an "anthropogene"(not an English word) is anything studied by such sciences.

Chapelle type of Neanderthal: a skeleton found in 1908 at La Chapelle-aux-Saints in France in association with Mousterian tools. Reflecting both the severe arthritis of the elderly specimen and contemporary beliefs that Neanderthals were primitive, it was reconstructed in a crooked and bent-legged pose. By the time Porshnev wrote a new reconstruction was available with an appearance like modern humans.

Holocene: the current geological epoch; an interglacial period beginning about 11,500 BP. Since there is no physical evidence for Neanderthal survival beyond 40,000 BP Porshnev's reference to the Holocene may be a mistake which would have been corrected in a finished version; however, see the paragraph (6) below.

Mousterian: a style of flint tools from the Middle Paleolithic, produced by Neanderthals and modern humans. Also widely distributed, they go as far back as 200,000 BP.

Neanthropine, Paleanthropine: from ne(o)anthropic, which refers to humans (*Homo sapiens*), and pale(o)anthropic, creatures such as Neanderthals known only from fossil remains.

Neolithic: the New Stone Age. Characterised by the introduction of farming, and later the use of pottery, it began as early as 10,000 BP in parts of the Middle East. Succeeded by the Copper and Bronze Ages.

Paleolithic: the Old Stone Age, applied to the very long period of human and pre-human evolution characterised by stone tools. Divided into Lower (the oldest, before

300,000 BP), Middle and Upper.

Pleistocene: a geological epoch also called the Ice Age. Beginning about 2,500,000 years ago, it is characterised by repeated glaciations. The Middle Pleistocene is reckoned to begin about 780,000 BP, and to end about 125,000 BP at the start of the interglacial before the current one; it was followed by the Upper Pleistocene and the Holocene.

Quaternary: the current geological period, covering the Pleistocene and Holocene.

The zoological term *higher primates* includes the family of great apes (also known as anthropoids), and the hominids. The phylogeny of the great apes is still far from being fully understood. There is a lot of discussion related to the taxonomy and phylogeny of the recent and fossil anthropoids, including the hominids' closest ancestors such as Australopithecus. The knowledge and study of the fossils of the hominid family is at this point still a young biological discipline. It is very likely that further studies will lead to substantial revision of the existing hypotheses.

Just one hundred years have passed since Haeckel made the assumption that it would be possible to discover a fossil intermediate between modern humans and extinct anthropoids. Seventy years have passed since this idea was definitively confirmed by Eugene Dubois' discovery in Java of the bones of Pithecanthropus. The idea of an intermediate type closer to the modern physical type came later – the so-called Neanderthal stage in the phylogeny of hominids. It wasn't until the 1920s and 1930s that most experts came to hold the view that *Homo neanderthalensis* was the immediate and direct evolutionary ancestor of *Homo sapiens.*

Although barely established as a scientific theory, this is certainly the potentially most interesting idea. But there has barely been time to start accumulating new knowledge and understanding regarding a number of problems associated with the various anthropogenic disciplines which, when solved, will probably require a complete re-evaluation of the total picture. Science is as always advancing, and should never be afraid to replace the concept that only yesterday was considered the most progressive, the most objective, even at the cost of destroying a lot of the established theories. The major recent changes in knowledge about the phylogeny of the hominid family can be summarised as follows:

1) In the past two decades, the development of Quaternary geology in the USSR (led by V. Gromov and others), as well as abroad, has gradually undermined various long-held archaeological and anthropological ideas, i.e. the *a priori* assumption that Neanderthals have changed into the modern physical

humanoid type (*Homo sapiens*). Immutable geological facts have shown that Neanderthals continued to live on in certain areas for a long time after *Homo sapiens* appeared in other areas. The current state of stratigraphical knowledge in Quaternary geology has led to a steady expansion of the overlapping time period where both species have co-existed. The time of the appearance of *Homo sapiens* has not changed in any significant way, but the actual time of the extinction of the Neanderthals has changed considerably and brought them well into the Holocene.

2) In the application of physical methods to determine the absolute age of archaeological finds and paleoanthropological remains, it has been necessary to make a revised list of hominid fossils of neanderthaloid type. To call osteological material Neanderthal it would not only have to have the necessary morphological features, but it should also be possible to attribute it to the Middle Pleistocene based on geological and paleontological occurrence or accompanied by Late Acheulean or Mousterian archaeological finds. Without this underpinning of the dating, even undisputed osteological material with neanderthaloid morphology remains uncertain. All sorts of found bones of neanderthaloid type conflict with these geological conditions, and are thus excluded from the Neanderthal series. A good part of these "silent" Neanderthal bones are well dated by physical methods, but are clearly not belonging to the Middle Pleistocene. But they have also broken the seemingly tight relationship between Neanderthals and the Acheulean-Mousterian Paleolithic. Previously they have been discarded as "atypical"or "atavistic" osteological material of neanderthaloid type, but of recent origin, but they are now back in the appropriate anatomical series.

3) During the past two decades, comparative anatomical analysis of fossil remains of human ancestors has gradually undermined the somewhat simplistic view of the "Neanderthal stage" anthropogenes. It turned out that the "classic" Neanderthals of the Chapelle type could be the descendants of neanthropines by extreme specialisation of some of their anatomical features. Now we have to consider that the neanthropine ancestors were the earliest form of the Neanderthals, and from these are descended on one hand the progressive new "sapient" forms, leading to *Homo sapiens*, and on the other hand the regressive form, characterised by "bestiality" of certain anatomical features. In other words, based on purely morphological terms the set of currently available data on Neanderthals (paleanthropines) shows a clear divergence.

4) These facts, as well as significant new hypotheses about the ecological niche, sources of food and food relationships of the fossil hominids (and the involuntary nature of their initial acquaintance with fire) force us to believe that the lifestyle and mentality of early Neanderthals were indicative of creatures who were on an animal level. Having considered the theoretical concept of "tools" that implicitly apply to the treated stones of the Lower Paleolithic, I have shown in several articles that the above view is consistent with the basic tenets of Marxism.

5) This new phylogeny of the hominid family should of course include the last split of the anthropogenes into two species or perhaps subspecies. This last phase began in a broad sense in the Upper Pleistocene in areas of the eastern Mediterranean. Neanthropines started to appear, forcing paleanthropines to withdraw and settle in other landscapes. In the end, this divergence caused the paleanthropic individuals to become increasingly rare. They still used treated stone tools, while *Homo sapiens*, who had assimilated various elements of their ancestors, evolved gradually and started forming societies. They did not assimilate the Neanderthals, so these relict forms had to adapt to the environment like highly organised animals to survive.

6) When did the last remnants of the now extinct hominids disappear? There is much to be said in favour of the fact that they died out very slowly, tenaciously holding on to their existence. Some findings as well as ancient monuments and epic literature suggest that it happened gradually from the Neolithic through the Copper Age, Bronze Age and beyond. However, it must be put aside when one considers the possibility of proving that these "living fossils" have survived in a few remote corners of the globe to the present day. A thorough review and analysis of all pertinent data is of course necessary before one can make such a fundamental change in the phylogeny of humanoids.

The question of the modern relict hominids should be considered primarily from the standpoint of zoology. However, many aspects of this area are unprecedented in the experience of most zoologists. One aspect is the similarity of the studied animal with modern humans. It is worth remembering that the study of the modern species of apes is still hampered by unsolved questions although it has been going on for at least two centuries. So are we dealing with animals or "wild" people? The special branch of biology that generally studies the family of hominids, that is to say anthropology, does not as such do field studies of wild animals. The biometric methods of anthropology can only be successfully applied to osteological material.

As is always the case in the history of science, some reputable scientists do not accept the hypothesis of a possible "co-existence" for any length of time, or the existence of paleanthropines and neanthropines (with the rapid progress of the first and regression into bestiality for the second), because they see this hypothesis clashing with the axioms of the science of anthropology. Thus G.F. Debetz, who is a professor of geology, expressed the opinion of himself and several zoologists regarding the continued existence of relict hominids into modern times with a joke borrowed from Chekhov: "It cannot be, because this can never happen." Since what we have in mind is incompatible with the axioms of an entire scientific discipline, this joke has a very deep meaning, and it should be taken very seriously. In other words, the arguments supporting new ideas have to be thought out carefully, and must undergo a comprehensive critical review before any final conclusions can be made.

This book deals with the complex but consistent shifts in the problems of relict hominid regression. Only one of the final chapters provides a brief overview of the problems regarding the extinction of paleanthropic Neanderthals. There are many reasons to believe that, for at least the last half century, people in our country as well as abroad have killed hundreds of specimens that could have provided a solid basis for scientific conclusions. Many of these captured or dead specimens were interpreted as “feral” gang members or fugitive criminals, or as a particularly savage kind of offenders, or perhaps mentally ill people who had left the company of others and had degraded to an animalistic state in the wild. They were all brought to administrative or medical institutions, where they had no voice, no name. It never occurred to anyone to study their subtle morphological differences from modern *Homo sapiens*. Systematic research in archives could only be undertaken with official permission. But such a government controlled project must of course be preceded by a thorough scientific discussion of prior knowledge and the advancement of theoretical principles. Without this one cannot of course expect any serious provision for field research.

Part One

Statement and Background

CHAPTER 1 • THE RIDDLE OF THE BIGFOOT

The Belgian scientist and doctor of zoology Bernard Heuvelmans has a special place among authors who have written about Bigfoot. It is not just because he, more than anyone else, has brought together all the available information from Western scholars on the yeti in the Himalayas [1]. It is not just because he was the first one to consider that the yeti actually exists, and to use comparative statistical methods to process the available data, which gave very convincing results. The most important thing is that Heuvelmans has shown that there is a very good opportunity for contemporary science to make this discovery official. He has explained the problem of Bigfoot on the background of zoological discoveries. Heuvelmans' great book *On the Track of Unknown Animals* is devoted to proving that the period of discovery of new species is far from over, though many people seem to think it is. There are still great possibilities for expanding our knowledge of the fauna of the globe.

"The great days of zoology are not done!" According to Heuvelmans we have only exhausted the possibilities when it comes to less rare and more easily observed species. We simply need new methods, and in a sense a new philosophy of zoology, to find other species. Previously zoology only got down to business when the office of a scientist received a skeleton or a stuffed animal skin. They did not pay any attention to simple rumours. Later, when zoology started to be taught in schools, extensive investigations would be made, or research concerning specific targets. As a general rule though the animals researched were more or less well known. In any case a variety of records about unknown animals do exist. They have not been prepared by a scientist with a scalpel, and they are not represented in any collection, so nobody takes them seriously. Why? There is nothing to indicate this information is false, so why can they not help the scientists?

In the beginning of the last century paleontologist Georges Cuvier had acted quite recklessly when he, in the introduction to his *Researches on quadruped fossil bones,* stated: "The hope of discovering new species of large quadrupeds is very small". But then in 1819 his student Lear sent him a sketch from India of an unknown tapir [2]. Until then zoologists had thought that tapirs were only American animals. It turned out that they had ignored the fact that ancient Chinese dictionaries and Japanese zoology books had descriptions of this animal. But in the last 100–150 years it has been clearly demonstrated, following a series of discoveries of large animals, that the arrogant words of Cuvier were wrong.

Thus the existence of the world's largest carnivore, the Far Eastern brown bear, only became known to Western European scholars in 1898, although of course Kamchatka, Sakhalin and Manchu hunters talked about it a lot. In 1900 we first learned of the existence of the largest land animal after the African elephant: the white rhino, reaching up to five metres in length and two metres in height. In 1901 the largest of the apes was discovered - the mountain gorilla, the male of which can reach 2 m 70 cm in height. Local residents had - of course - told stories about these giants for a long time.

Until 1912, zoologists who were far too confident of their own knowledge and methods believed that the Komodo dragon people talked about was a myth. That year it turned out to be a real giant lizard, in fact the largest of all the lizards on earth. Let us take another example – a well known one – the discovery of the okapi.*

Local people in Rhodesia had for many years told about the existence of some kind of predator, which they called, if translated literally, "leopard-hyena", but European science did not recognise the reality behind these stories until 1926, when this new species [3] was named "king cheetah." The English zoologist Pocock who described it observes: "It is surprising that such a large animal so different from other types remained unknown for so long". It is not surprising, according to Heuvelmans. He feels this will continue until the scientists stop regarding the stories of local residents, hunters and connoisseurs of the region as absurd inventions.

Heuvelmans' book *On the Track of Unknown Animals* is a severe accusation of the traditional conservatism in zoology. The zoologists were for example adamant that the so-called Bermuda petrel had been exterminated at the beginning of the seventeenth century. And yet recently, in 1951, a freshly inhabited nest of this bird was found on one of the islands around Bermuda. So Heuvelmans compiled an extensive catalogue and gave preliminary descriptions of a large number of amazing animals, birds and sea creatures worth further investigation. The discovery of new animals need not be an accident, says Heuvelmans. We have not yet fully explored the planet, we do not know much about it, and many areas have in fact never been visited.

In the book *On the Track of Unknown Animals* there is also a large chapter with a summary and analysis of preliminary data regarding the amazing great ape that lives in the highlands of the Himalayas and is known as the yeti. With such a broad background of data this problem immediately changes from being something exceptional to just another zoological problem. The zoologists should be blamed for not trying to solve it. They should not blame the problem.

Yet the question remains: Is there a "snowman" and which scientific pathways will lead to the validation and proof of its existence?

* *Translators's note: some text is missing here.*

On the latter point the existing opinions can be divided into two groups. The vast majority of people think that the mystery of Bigfoot will be solved in one single sensational act. Someone will catch him or kill him, and this will start the scientific study of this creature. All that has happened before this particular incident is of absolutely no scientific value. Others however, a minority consisting of almost all the experts in this matter, believe that the capture or killing of one of these animals will only serve as a confirmation, the final stage of a large preliminary study.

You think you can catch Bigfoot sitting at a desk. You can read books and letters till the end of time. No matter how much oral or literary data you collect, it will not bring you one step closer to solving the question! Many sceptics make jokes like this. But it is even easier to refute and ridicule the idea of a sensational discovery that supposedly can solve large, complex scientific problems. Science knows only too well, that a spark is of no use if you haven't got a barrel of gunpowder. First of all, you need a reasonable working hypothesis, some sort of preliminary idea. Without this, all this talk of sensational discovery can only damage a serious study of the Bigfoot problem. It brings frustration to everybody, brings out expeditions consisting of inexperienced persons without any specialised training and knowledge, and creates fantastic and largely untrue stories in newspapers.

My best advice to these people is "Calm down!". A "first" is nothing if it is not the first of many.

People who are impatient and criticise the slowness of scientists are not aware of how many times in the past people have tried to do what they naively think will solve everything. Find! Catch! Kill!

In later chapters you will find, for example, the description of how geologist and experienced hunter M. A. Stronin in 1948 in the eastern Tien Shan was aiming at a strange creature at 100 m, but did not dare shoot. "All I could think of was what kind of creature this was." He clearly saw that this was neither man nor bear. He never told anybody about it. If he had a vision of a "snowman", he could very well be fired from his job.

In 1941 Doctor V.S. Karapetyan clearly saw a humanoid creature overgrown with hair in the mountains. He did not pay much attention to it at the time. "But," he says, "I was only seventeen years old back then, and I only wondered what it was. Now, having read something about Bigfoot, I start to think that my observation might be useful for the scientists who are interested in this issue. If only I had known something about the question then…"

In 1939, during the battle at the Khalkhyn Gol River in Mongolia, Soviet officer G. N. Kalpashnikov examined the bodies of two hairy humanoids that had been killed by mistake. "At that time, I knew absolutely nothing about 'Bigfoot', and I still do not claim that those beings that were killed were in fact 'snowmen'. But just like others I wondered who they were, but at the time, in the midst of fighting, we were busy with military affairs, and we were not up to the study of those animals, which are now of interest to scientists."

In 1934 geologist B.M. Zdorikov, who was working in the Pamir Mountains, and his guide came across a hairy long-legged animal sleeping in some tall grass. It was definitely not a bear. The local people told him something about those creatures. "It is only now I recognise some of the features of the yeti as it has been described by the Sherpa (Nepalese mountaineers). At the time I had not heard of the Himalayan yeti, and thought the stories I heard in the Pamir Mountains were just part of some fantastic local fiction. When I finally met the beast myself, I could not believe my eyes, nor could I believe the local stories about humanoid creatures living in the wild mountains."

In 1925 Greek photographer and geologist N.A. Tombazi, who was travelling in the Himalayas, saw a human-like animal at a distance of about 200 m and studied its tracks, but he said that he was "just not in a position to express any definite opinion on the matter". Of course, had it just been fairy tales as told by his Sherpa porters, this educated gentleman would not have taken them seriously.

I could go on listing similar sightings, but there is no need. These are only to illustrate the general structure. Every time the observers did not know what to do, and could not explain their observation. They had to resort to some form of home-made explanation or keep on disbelieving their own experience.

So are there no cases where this creature has been caught? Yes, there are, and they will presented below. There are several stories of local people who have caught and even domesticated the animals. Here for instance are two reports from two highly cultured people, who at different times and in different places bumped into specimens captured by local people, and described their experience in interesting detail. 35–40 years ago zoologist, hunter and author N. Baikov, accompanied by a hunter named Boboshin, came across the lonely hut of a Manchu hunter in the remote foothills of the Manchurian taiga, and met the most amazing humanoid beast.

Baikov published his description long before the publication of any reports of Bigfoot. In 1954 a prominent Chinese historian, Professor Hou Wai-lu, saw a similar creature that had been caught and tamed by local people in the mountains of Qinling Shan. It did not have the capacity to speak. Professor Hou Wai-lu does not think that this creature is the same as the "snowman". But on the other hand the Chinese historian had heard nothing about the nature of Bigfoot. As you can see, in both cases the observers came in contact with captured specimens. They did not know each other, and did not reach any conclusion. Each had to explain their sightings as best they could. In these cases the spark did not ignite a scientific explosion.

The main obstacle to the solving of the puzzle of Bigfoot was apparently the lack of available consistent data to the individual observers and researchers working in Asia. None of them knew that information about these creatures was available in other areas, and so were not able to perform one of the most basic scientific exercises – the comparison. And as a result, their observations just faded away and were forgotten. It is only very recently that some reports of Bigfoot have become widely known. Now it is less likely that a casual observer would not believe it, or could not remember having heard of anything like that. But bringing together a large number of observations and evidence provides solid ground for scientific generalisations and conclusions.

The human mind does not like a blank, so it has tried to form some sort of explanation or at least give a name to the observations. Some of them have lived on in traditional folk interpretation. In quite a number of cases the wild hairy human-like creatures have been interpreted as people who have become wild because they have stayed for a long time in the solitude of the mountains and deserts of Asia. According to storytellers and local authorities, they were the descendants of people who had been forced to flee and hide from religious persecution, people isolated from human society, the lepers, the descendants of unwanted tribes or unwanted childbirths, ousted neighbours, or in some cases not even the descendants, but "feral" or "hairy" criminals who have hidden themselves from others. In other cases, they have been described as cases of abnormalities, atavism, deformities, mental illness or similar things. Finally, a very large number of sightings have been treated as belonging to the realm of the supernatural. Those who believe in evil spirits suggest that observers have met with them or creatures of superstition and legends.

People may be educated, but if they believe in ghosts they will include all kinds of messages and stories in their beliefs. So consequently most of these stories are relegated to the field of folklore, no matter how much the witness swears they had seen the creature or studied its tracks or heard its screams. If these stories are recorded and studied at all, it is usually done by ethnographers and folklorists who study the myths and legends of the supernatural, stories born of popular imagination. So until very recently many of these stories just lay scattered under different headings, in different departments under different disciplines.

Science needs to be able to compare things. The scientific study of the Bigfoot problem only started when enough material had been accumulated for researchers to be able to compare the various descriptions. At first it only came from the Himalaya region. The successful approach of the comparative method can be demonstrated by the articles of Bernard Heuvelmans. After having studied the various accounts available to him, he started to quantify them in various ways. Here are some of his conclusions. In nine out of eighteen major reports the animal is described as man-like, according to seven reports it looks more like a monkey, and according to two reports it is a mixture of both. In thirteen descriptions the creature was two-legged, three said it was a biped that could move around on all fours. In one description it only moved on all fours, although it was only seen climbing among some rocks. Fourteen reports describe hair on the body of the creature. In six cases the witness stresses that there was no hair on the face. Heuvelmans has also grouped the reports according to the colour of the fur and a number of other features. Some of these are constant, others show a greater degree of variability. But in general, there is no contradiction in the various descriptions according to Heuvelmans. Most features converge, and the deviations you do see are not particularly significant. The evidence in the majority of these cases is also quite independent, coming from people belonging to different nationalities, different levels of education etc.

We shall not discuss the findings of Heuvelmans in detail. We only need them as an illustration of the comparative method. Another useful table produced by Heuvelmans was the location of all the information in chronological order. As for the topographic data, in particular data on the observations of tracks within the Himalayas and the Karakorum, this was made by the French geology professor P. Bordet and others.

Although foreign researchers have gathered a certain amount of data from the Himalayas and the Karakorum, comparing these data has only given limited evidence for the reality of the Bigfoot. For cautious authors like S.V. Obruchev [4], the information had to be classified as "reliable" or "unreliable", but this necessarily involved a certain subjective standard. This method is of limited value, so we still try to find objective criteria so as to expand the field of application of the comparative method.

In January 1958 the Academy of Sciences of the USSR established a commission to examine the issue of yeti and Bigfoot. In 1959 it was converted into a public research organisation, bringing together various experts. At the same time the Academy of Sciences continued work on the gathering and synthesis of information on the question of Bigfoot. During this development the author of this book, first as vice-president, then as president of the commission and the one responsible for the relevant topic in the Commission for the Protection of Nature in the USSR, was guided by the possibility of expanding the application of the comparative method. On one hand they were careful to draw a parallel with some of the findings of studies of the ecology of fossil hominids. On the other hand, and above all, it was the attraction of these creatures, according to the description, being similar to the "snowman" from other mountain areas of Asia.

However, there is always the risk of making an error. These beings may be similar, but that does not necessarily mean they are identical, nor even related. But now it seems that the danger of making such a mistake is a thing of the past. The comparative method proceeds to be fruitful, and that has led to a new stage in the development of the Bigfoot problem. In an article in the newspaper *Pravda* in July 1958, the author wrote about the yeti creature, which for a long time has been described under the name *almas* in the territory of the Mongolian People's Republic. A Paris newspaper commented that this was the first time the question of the yeti had been discussed on a scientific basis. What was originally just a hunch had paid off, and greatly expanded the field of research. This breakthrough was originally based on a fairly narrow portion of the collected scientific information (only from the southern slopes of the Himalayas and surrounding ranges). Now, instead of just the information gathered in the Himalayas by Shipton and Tilman, Stonor and Izzard, Heuvelmans and Bordet, there are also the findings of many other naturalists, geographers and ethnographers, collected at different times and in different geographical areas of Asia. This has not only increased the possibility of using the comparative method, but also the reliability of the results.

Where Heuvelmans only compared eighteen different sightings, we now have more than a thousand. A significant part of them has been published as a source of raw material for researchers in the form of small collections: *Information Materials of the Study Commission on the Snowman*, ed. B. F. Piston and A. A. Shmakova (1–4, 1958 –1959).

The compilers of all the editions of the "Information Materials" have stated that the reliability of the sightings and documents included in the collections are extremely varied. Regardless of their attempts to assess each sighting individually, this is the result whether you just study the files of Bigfoot or the entire set of source documents. The researchers have had the possibility to criticise and refute the data. But now it is the first time everybody has had the opportunity to access the actual data.

There are four issues of "Information Materials" so far. This is the main result of the first two years of our research into the question of Bigfoot. Further collected materials (issues 5 and 6) are still waiting to be published [5]. The gathering of evidence and sightings within the USSR was largely made possible through the help of the Soviet press. People reacted when the newspapers published notes on the problem of Bigfoot from time to time. The letters that came to the offices of newspapers and magazines focused on the commission set up to study the question of Bigfoot. They often provided very valuable information, and there was no reason to believe it was unreliable gossip about the search for Bigfoot. Generally, it is very easy for a specialist to tell if a message is inspired by the press or fictional rumours. After all, the ingenuity of the human mind is not limitless. The overwhelming mass of correspondence as well as oral communications indicated a sense of responsibility and duty by the witnesses. There was a complete lack of interest in the actual publication or in any kind of fame. The authors of the letters usually motivated their letters like this: "I consider it my duty to report to the science…", "I consider it my civic duty…", "maybe my humble message is useful to our scientists…" Along with this influx of voluntary information, the Commission has gradually established a network of correspondents in various areas – local scholars, local historians, hunters – who interview the local population. Their selfless assistance has been a significant help during the study. Finally, a certain amount of work has been done by special expeditions to some mountainous regions of the USSR. As for the foreign media, the Commission has received information from a number of countries from the professionals involved in the study of the problem of Bigfoot. Their cooperation has made it possible to make our reports as complete as they are.

So indeed, every effort was made to ensure that the role of suggestion and tipping played a minimal role. It would also be difficult to explain why a significant number of people, without getting any benefit out of this, would dare to cheat a handful of scientists or even the general public. It would be rather reckless to postulate all the informants are wrong. At the same time, if our informants simply invented the sightings, they would all be very different. But in actual fact there is no significant difference. Another significant proof is the discovery of a fairly large amount of old records and material, for the most part long-forgotten, which, as it turned out, do not disagree with the newer sightings, but on the contrary are basically the same.

The collected material may be subjected to a wide range of comparative and quantitative (if not statistical) treatments. On this basis, the results of the comparative method have a fairly high degree of certainty.

The information for the most part comes from territories very distant from each other, from people that are widely separated culturally as well as linguistically. It has been recorded in different historical periods, and has been largely neglected. In some cases the stories have been repeated by representatives of the local population, but also told to visiting people who did not belong to the local population and did not have time to get intelligence from the local population. This can only be interpreted as proof that this is objective information based on the observation of existing natural phenomena. The same can be said regarding the overwhelming majority of the oral information. Real physical data (tracks, handprints, hairs etc.) although few in number, as well as oral description of many morphological and ecological aspects, has given us a clear biological sense of the phenomena when the variety of information is synthesised by a zoologist. It is also important that the information collected from the oral accounts of Bigfoot, despite all their diversity, as a rule do not refute each other. The differences between the individual parts is well within the biological understanding of intraspecific variation. In this case, the resultant understanding of the morphology and biology of Bigfoot, as we shall see, is quite rational in the eyes of a naturalist.

This is the scientific basis of much increased confidence in the reality of Bigfoot.

Any of the sightings in the "Information Materials" can, when considered by itself, be questioned by a sceptic, but no one who has read all four issues has been able to explain the unity and consistency of the material taken as a whole. Thus, after having spent a five-year period of orientation in the vague mass of a variety of information, we are more and more confident of the reality of the facts beneath it all.

The biological interpretation of Bigfoot is becoming more and more sound, whereas the attempt to interpret it in a mythological context is an evident failure. The idea that all reports of Bigfoot, "wild man" and so on, exclusively come from the field of popular fantasy, is contrary to the science of mythology and folklore. It would be sad if the current fields of ethnography and mythology in the second half of the twentieth century did not take the question of folk beliefs into account. To make a major discovery in the field of folk belief is no less difficult than in any other branch of science. It is even harder to overturn its general root of how culture and ethnicity relate to each other and how legendary stories and images are distributed. The idea that Bigfoot represents a single cycle of legends, common among many Asian people belonging to different cultural, historical, linguistic and religious groups, and that this cycle of legends somehow remained unnoticed by folklorists, would truly be a revolution in the study of folklore.

On the other hand, compared to the mythological versions are the realistic results based on the quantitative analysis of the collected mass of primary documentation. If we do it the mythological way, the more we collect of short stories recorded in different geographical areas, the more our material will be expected to spread out and the image dissipate. In the end all signs will cancel each other out. In this case we get the opposite picture. It would seem, for example, that the large number of references to meeting with a Bigfoot in the night raises a suspicion of similarities with stories of "evil spirits". But the quantitative analysis has shown a much greater number of meetings in the morning and evening, and only a negligibly small number of references to the daytime. If we were talking about fiction, why should the narrator specify these times of the day, and not just let it be during the mysterious darkness of the night? The point is though, that the timing of the sightings to a biologist suggests a twilight-night daily life cycle, very similar to the daily activity of many species found in the mountain fauna.

In general, when you process the mass of material about the yeti, there is a high degree of specificity and constancy of the described biological characteristics and properties. All in all a good argument against the mythological version.

How could anyone entertain the idea, that all this is nothing but myths misunderstood by the local people, when the inhabitants of the different nations in Asia have described similar anatomical and biological details? The description of the wild man as found in the stories of the Kazakhs is, apart from a few minor differences, consistent with the descriptions given by the Sherpas in the Himalayas, from the Manchus and Mongols, from the Tajik and Kyrgyz. In 1914 Russian zoologist V.A. Khakhlov gathered basically the same information in the Tien Shan mountains, but in a more accurate and complete version, than the English zoologist Charles Stonor did forty years later in 1954 in the Himalayas. That alone should be enough to drop the mythological version. We will however return to that in the relevant section.

Any attempt to explain the amount of information about the "snowman" as a reference to some sort of "devil", "brownie" or ghost is basically nothing but an attempt to brush off the problem. What use is science, as for instance in the foreword to Charles Stonor's book *The Sherpa and the Snowman* [6], as written by anthropologist G.F. Debetz, if the author is comparing the yeti and the "devil", and completely ignores the existence of scientific literature. And there is another problem as well. The recording of folk legends and beliefs made by ethnographers is subject to certain rules. Instead of using all of this seriously Debetz refers to his own hazy memories: "Can I take the stories of Sherpas as indisputable evidence of the existence of the yeti? No, I cannot". In 1929 the author of these lines was in the upper reaches of the Ilim in the Irkutsk region. Today there is a railway, but back then there was not even a school, and not a single literate person among the local natives. The author was often told stories about the devils, and with details in no way inferior to the Sherpas histories of the yeti. One cannot base science on such home-grown "memoirs". How can one now check what was actually heard by Debetz? Was it stories about the devil or perhaps about the real analogue yeti, or had he heard different stories in other places?

The common belief in an "evil force" among many different people can be explained by ignorance. But even in the case of scholars who study the specific form of personal beliefs problems can arise. They would not do well in natural science, nor would it be possible for a biologist to do well in folklore.

For example, zoologist N. Vereshchagin [7] said "the evidence for Bigfoot belongs in the department of goblins and witches" - although he was not an expert in this department, and therefore does not know how folklorists actually work. The author knows nothing of the data and the current state of the question of Bigfoot. For example, he criticises Stonor, Izzard and Soviet researchers on this subject based on "the current level of biological science," for claiming that there is a primate habitat in the area of permanent snow and ice, but Stonor and Izzard completely rejected this assumption in 1954, and now no author in the world makes similar claims. Vereshchagin confused without exception the local names of the creatures, for instance by confusing the term "almas" with the name of an adventure film and claiming that such creatures do not exist in Mongolia, Tibet, the Pamirs and the Caucasus. The author even provides "reliable" information about some fictional groups, allegedly searching for Bigfoot in the summer of 1959 in the upper reaches of the Samur and Sudak rivers. In short, the article "The myth of Bigfoot" offers not a single bit of knowledge on the subject. It is hard to understand just why serious scientists are suddenly starting to write about a subject they have never studied.

One can cite a few other examples. In the pages of the Ukrainian magazine *Science and Life* Professor I.G. Pidoplichko has devoted an article to the yeti as a figment of people's imagination. This article is just one of many, and one must wonder why the author would write about stories that were completely unfamiliar to him. Geographer Professor E.M. Murzaev published another article "The modern myth of Bigfoot". Why "modern" asks the reader? In Murzaev's opinion, based on information from the ethnographer A.Z. Rosenfeld, “Bigfoot, *gul-biyavan, yabalyk-adam, alamas* (sic!) - all these heroes of folk tales are nothing more than versions of a shape-shifting bear.” But in contrast to the above-mentioned articles, Murzaev's writings are not just amateurish, they are misleading: the author states that at the request of “a Moscow scientist” (referring to B.F. Porshnev), during an expedition in 1960 to Xinjiang, in the Kunlun Mountains and elsewhere, a “thorough survey” of the local population about the Bigfoot was made, and the result was negative. Members of the expedition actually reported the exact opposite.

Why did Murzaev not write that a member of the expedition, Professor B.A. Y[*], wrote in his field diary the story told by the resident of a village of Tashkurgan about a person who had seen the corpse of a “wild man"? It is possible that Murzaev did not know that B.A. Y gave this record as well as several others to the Commission, and informed them that the leader of the expedition, Murzaev, had specifically forbidden him to study these data. In passing Murzaev writes this remark: “Some hunters still maintained, that they would meet *yabalyk-adam* in the border area between the Soviet Union, China and India. He would always be alone, and would not react even though people threw stones at him.” Murzaev clearly dismisses this as a joke, but as can be seen from the articles, Murzaev did not bother to record the stories of the local residents in his field diary. The study of myths and folklore requires documentation too. Instead we now have an unclear folk story about a fight with the Bigfoot.

Still, the question of Bigfoot is not a done deal yet in modern science. Everything that has been done, everything stated in this book, is nothing but a search for the right solution. Our attempt at a systematic examination and synthesis of the existing data should be viewed as a way of constructing a working hypothesis. We are still at the very beginning of the path to solving the puzzle. No matter how large the amount of collected material appears compared to yesterday, tomorrow it will probably look negligible.

For the time being, we will be in the same position as the astronomers Adams and Le Verrier, who in 1846 calculated both the point in the sky where you would expect the planet Neptune to be and its mass. Le Verrier wrote: “I see it with my mental eye, I feel the tentacles of its mathematical analysis as clearly as if it were in front of my eyes ...” Two weeks later, Galle discovered the planet. The difference, however, is that no one was looking for the planet Neptune before it was predicted. We, on the other hand, have a set of observations, but no general consensus.

In 1961 there were several attempts in the foreign press to summarise the results of previous studies and the information about the mystery of Bigfoot. The weakest of

* *Translator's note: Y is not a Russian name. However it is common in Russian to say Mr. Y or Professor Y, when in English we would say Mr. X or Professor X. Porshnev might have been hiding the professor's identity, or had simply forgotten and was planning to come back later and rectify the omission.*

these efforts was provided by the French anthropologist Professor Henri Vallois: "What can we say about the yeti, weighing all the facts and stories?". This attempt is not only based on completely worthless information, but in no small measure also an example of unfair misinformation in the first place. Vallois had full confidence in the material of the Hillary expedition, and this has now been exposed as a lie, a deliberate attempt to mislead the world of science [8]. Also, Vallois seems to know only the two first editions of our information material, and nothing of the further releases. Can this truly be called "well-balanced" information? This article by Prof. Vallois has nothing to do with science. The scientific article "Current status of Bigfoot" [9] of the prominent British primatologist Dr. Osman Hill is far more interesting. Unfortunately, the author knows very little about Soviet research, but the most important data from the Himalayas have been subjected to careful and thoughtful analysis.

The book by the English zoologist Odette Tchernine [10] is also of great interest. This is an excellent summary of all the data in the literature that the author was able to collect. There is a lot of new material relating to the history of the study of this subject. The author, like Osman Hill, is leaning towards recognition of the Bigfoot as a biological reality – a hitherto unknown kind of great ape (the orangutan included). The author draws some very interesting conclusions regarding the biogeographic conditions for the continued existence of this animal and possible reintroduction in a vast expanse of uninhabited regions of Asia and America. Tchernine has greatly expanded the scientific credibility of this subject, and has drawn the interest of American and Soviet experts. Unfortunately, many Soviet studies were largely unknown to her. For an author writing in broad scientific terms, it is unfortunate that much information from the Himalayas is restricted. There is not enough information available.

The first scientific study of this kind was published in 1961 by the famous American biologist Ivan T. Sanderson: *Abominable Snowman, Legend Come to Life. The Story of Sub-Humans on Five Continents from the Early Ice Age until Today* [11]. This extensive and thorough work has made a great impression. Here the author has made an attempt to review and organise a huge amount of material on the issue of Bigfoot. Unfortunately, he pushed the "classic" Himalayan data too far into the background, and did not give them a lot of space. It is very controversial whether Sanderson did right when he highlighted the data from Canada and North America to such a degree. He has set aside six chapters at the beginning of the book for it. But he has also explained that it was a conscious desire to, as it were, destroy the "Himalaya-centrism" on this subject. The author used the comparative method throughout the book. I am pleased to note, to a certain degree, our research gave a lot of impetus to his work. As Sanderson said: "This Soviet activity shed an entirely new light on the whole issue, and raised it as a whole to such a high level that Western academics were almost forced to radically change their position in relation to it." The point here is not only the wide geographical scope of the research, the broad parallels and comparisons, but also the moving of the whole problem from the biological realm of the anthropoids to the hominids. This will be discussed in detail later on.

The publication of Sanderson's monograph indicated the start of a whole new stage in the development of the study of Bigfoot. Our different opinions on certain issues will be discussed below. But we certainly concur with him in the understanding of the main tasks of today: in order to be able to move forward, you need to compile, synthesise and organise the gigantic preliminary material available, and produce a meaningful model of the studied object, a working idea, that science can evaluate. Anybody who reads Sanderson's bright and impressive book will feel the breath of this new stage of international work – as the transition from adolescence to youth is underway, the notion of maturity must be preserved for the next phase, which is now easier to predict than it was a year ago.

This book was more or less completed by the author in July 1960. It is evidence of the acute need for such a general work. And it is quite remarkable that both of our books have been written more or less simultaneously. Our books are not just a reflection of the stage of the research of today. They are also a necessary condition for the transition to the knowledge of tomorrow.

CHAPTER 2 • FROM THE PAGES OF ANCIENT TEXTS

In one of the most ancient written texts, the Babylonian "Epic of Gilgamesh" from the third millennium BC (the origins of which date back to the fourth millennium BC), we find the description of a strange humanoid creature, Ea-bani or Enkidu. The epic story tells of the fighting and later friendship between King Gilgamesh and Enkidu, a wild man created by the gods and sent down to stop Gilgamesh from oppressing the people of Uruk. After a titanic wrestling match, Gilgamesh and Enkidu become firm friends and work and fight together to protect Uruk. Among other things Gilgamesh and Enkidu fight together in the mountains of Lebanon to protect Uruk from a race of hairy mysterious beings that live in the wilderness there.

At first a shepherd and hunter from the Euphrates Valley describes a flock of animals arriving to drink water near where the people live. These beings come from the steppes and the mountains. They are wild human-like animals, the Ea-bath, "created by the goddess Aruru... Ea-bath has no clothes, all of his body is covered in hair, and the hair on his head is like the hair of a woman, yellow like the strands of a sheaf." These strange beings do not communicate with people. They live on plants like the gazelles, and come to ponds and springs to drink with the animals. The wild man-beasts live in close contact with the herds of wild herbivores. They suck their milk, wander along with them on the steppes and in the mountains, and protect them from lions, leopards and wolves. Ea-bath know how to fight one and tame the other, and they know how to protect the wild herbivores from man. The shepherd and hunter complains that if he digs a pit to trap the animals, the Ea-bath will pull them out of the traps, or even pull them out of his hands. This "wild man" with the hairy body knows nothing of the human way: he never cuts his flowing hair, and does not wear clothes. He would not eat bread or drink beer, his mind was not like a human, and he had no "understanding" (speech). But he could run faster than any man, had "stronger bones" than even the legendary Gilgamesh, was immensely strong, and grew up among the wild beasts.

There are a lot of important ecological references to the animal. It came "from the mountains" or "originated in the mountains", although in some cases it says "born in the desert and nurtured by the mountains", or of "a product of the steppe". In the description of the death of Gilgamesh, one finds a long list of the local fauna: antelope and onager from the desert and in distant pastures. There are also bears, hyenas, leopards and tigers, goats and lynx, lions, deer and chamois, steppe cattle and other beasts, panthers of the desert. It is also apparent that Enkidu was in fact one of these hairy beings, and that he came from "the mountains of Lebanon", which today are the mountains of the eastern Mediterranean, and that he spent his life in these mountains and the Syrian desert before turning up at the banks of the Euphrates.

The shepherd-hunter who first discovered the wild man drove his cattle home in a hurry. He was frightened, and his face was numb and stiff like the face of a dead man. He told his father about the creature, that "nobody dares to approach". The old and wise father decides to try and tame the beast with a female whore, hoping that the difference between the wild man and the woman was small enough for him to direct his sexual instincts towards her. His plan finally met with success. According to the epic a magical transformation took place in the wild beast after seven days with the whore, and he turned into a human animal. There are a few older fragments of the legend that describe the gradual taming of the beast, how it learned to eat human food, and became like a guard dog. Now the sheep herders could sleep at night, because the wild man would be awake and ward off predators. How long he remained like that is unknown.

The epic also describes how King Gilgamesh travels in the mountains of Lebanon to fight the Humbaba (or Huwawa) that live in the cedar forests. There is also a description of the loud scream of the Humbaba. This is also the point where we learn that Enkidu used to be one of them. The elders of Uruk bid Gilgamesh farewell and tell him: "Let Enkidu go before you, and those who will walk the pathways, he knows the forest ways and the habits of the Huwawa." And Enkidu himself says: "my friend, when I lived in the wild I roamed the roads in the mountains with the wild beasts... Fear not, trust me, I know this place well, and I know the ways of the Huwawa." Gilgamesh is successful. Not only does he kill the Huwawa, but also "seven fears" that roamed the same forests. To them it was a sacred trust to destroy the horrible creatures that lived in these impenetrable forests.

These are the oldest preserved stories in the memory of humanity, and in time they have fused with newer myths of hairy manlike animals. It is worth noting in passing that these mountain areas in the eastern Mediterranean have produced many important paleoanthropological discoveries, among them many transitional forms from ape to people of the modern physical type. Here the scientists are willing to consider the transformation of Neanderthal man.

The name Ea-bani has attracted the attention of researchers. The name is essentially the same we hear now five thousand years later from Iranian-speaking people such as the Tajiks, who calls similar creatures *biabani*, *yavoni*, *yavoi* etc. Similar names can be found among the Azerbaijani – *biaban*, *guleybani*.

Another of mankind's oldest books, the Bible, also contains a half-erased description of similar creatures. A detailed description has been made by the Hebrew scholar Rabbi Yonah ibn Aaron, but we will not go into details here. It has already been mentioned in the book by Ivan Sanderson. These creatures are known by names like *giborim*, *seir*, "the shadow," "the hairy ones ", "the destroyers", "the hunting creatures" and others. They are different from humans, have red hair on the body, a powerful smell, and a loud voice. According to Yonah ibn Aaron their hairiness indicates that they were hominids, and therefore capable of interbreeding with people of a modern physical type if they were capable of communicating with them. One can possibly also interpret, for example, Esau and his sons in a whole new way. They had long arms; their body was covered with red hair. They were no more than 4.5 feet in height, with short but straight legs. Their elbows, neck and feet were unusually wide for a normal person.

The hairy ones lived on the Sinai Peninsula and in the south of Egypt. During the stay in Egypt, the Jews were often closely associated with the "hairy ones" (perhaps they had domesticated them like animals?). They attacked them and kidnapped their children, but they tried to hide in pits covered with branches or leaves. When the Jews were evicted from Egypt, they started becoming more violent towards the "hairy ones". Yonah ibn Aaron has shown that the description of the ritual sacrifice of two goats actually describes one goat, which was slaughtered, but the other one was originally a *seir*, but this name has later been changed to *seir izim* meaning hairy goat. In the end this animal was not sacrificed, but was allowed to go back into the wild, free for all human sin.

The Greek author Ctesias, a former doctor at the court of the Persian king Artaxerxes II, wrote that in the mountains of India, i.e. the Himalayas, there were wild human-like animals. Other ancient naturalists, and later European authors, borrowed this description from Ctesias. I wonder if we are searching for the same animals now two thousand years later, when we are looking for the yeti or Bigfoot. Apparently other people knew about these creatures in ancient times. On the podium of the amphitheatre in Pompeii there are paintings of five scenes of fighting wild animals; there are tigers and apes, but also human-like creatures. Some have suggested they are gorillas, but I have my doubts about this possibility. Some scenes look like they are happening in a circus, and the images are quite clear. Pliny and the other ancient writers who knew the details of these human-like animals all ascribed them to the mountains of Asia, and this is not where we find gorillas.

Earlier authors summarised any information about wild and feral humans, even some that to our eyes is very curious. One article was published in 1731 in the historical, genealogical and geographical notes to the *St. Petersburg Vedomosti.* [Vedomosti being a newspaper] It said: "Ordinary people still believe that another kind of wild people have free run in the forest. This suggests they are nothing more than pagan fables about satyrs and Silvania. These were portrayed almost the same way as contemporary artists paint the devil. They live in caves and other secret places. And should they capture a woman, they would not let her go, until they have finished their fornication with her (Heraclitus of Tyre). They act and move in a disorderly way, but they never say anything. They call them satyrs, and many of them are to be found in India (Pliny). As Sulla returned from the war with Mithridates, they brought him a satyr, which was caught near Dirrehma (Additions to Libya Titus). "

Medieval European travellers in Asia make several references to "wild people" or humanoid animals living in mountains and desert areas. One of them was Giovanni del Plano Carpini, who in the middle of the thirteenth century travelled to visit the Mongols at the request of Pope Innocent IV. In his *History of the Mongols*, he wrote: "To the south of the city of Hanyla (probably modern day Imil, slightly to the west of Lake Kizilbash) is a large desert. And here the wild (forest) people are said to live. They do not have any form of speech." And Plano Carpini writes about a series of implausible rumours about the absence of joints in the legs of these creatures. "Apparently they are able to run very quickly away, when the Tartars are pursuing them. If the Tartars hit them with arrows, they will press a handful of grass to the wound."

Sometime around 1396, a German (Bavarian) by the name of Johann Schiltberger was captured by the Turks, and ended up spending over 30 years in Asia. He visited Mongolia and Siberia and met Timur and visited at the court of the Shah of Persia. He later wrote that "I visited this country, and saw everything with my own eyes."

When he returned to Bavaria in 1427, Schiltberger wrote a very famous travel book. The most interesting part is some of the observations he made in "the land of Siberia", which however may in fact mean Mongolia. He writes that "in this country there is a mountain of great height called Arbuss". There is a mountain of the same name known to the Mongols at the south-western border of the Gobi Desert, although it could also be the Altai area. "People living there think that it is a mountain of the desert, tall enough to reach to the heavens. No-one can go through the desert and live there, because it is full of wild animals and reptiles. On the mountain above the desert there are wild people who do not have permanent homes. Their bodies are covered with hairs, except the hands and the face. They wander in the mountains like other animals. They eat leaves and grass and everything else they need. The lord of the country sent to Khan Edigi a man and a woman who were caught on the mountain. And also three wild horses. The horses living in these mountains are the size of a donkey. There are also many animals that are not found in German lands, and so I cannot name them…"

A very important part of this story is the reference to the capture of small wild horses. This seems to be the first mention in European literature of the Przewalski horse. The authenticity of this animal indirectly confirms the existence of the other animals found in the mountains and seen by Schiltberger, the wild and hairy humanoid animals.

In Central Asia there are indeed lots of rumours of the existence of wild men. Yet zoologists often doubt these, but not stories about wild horses or wild camels. Here is what we know from the text of the Kyrgyz national epic "Manas", which comes from the middle of the ninth century. It has been carefully passed on by word of mouth from generation to generation. According to the tale, Manas's troops came out of the

Minusinsk steppes, crossed the Orkhon River and moved on to China. Interestingly enough, in the northern part of the Xinjiang province in China, there is now a river city called Manas. Anyway – Manas addresses his followers with these words:

'And do you know, there are
miracles no one has thought about
on the great path ahead.
What have we yet to see?
Who has heard of the wild camel.
Who has heard of the wild people?'

(Translated by S. Lipkin)

Numerous written records in China and Mongolia and from other people in Asia contain information about the ancient anthropoid wild creatures. The work of collecting all this information has just begun. We have for instance long but fragmented descriptions from experts in ancient Chinese writings and chronicles. The Chinese scientists Pei Wen-chung, Wu Liu Kan and Cho Min-tsyna have written an article about the "Mysterious Bigfoot", which was published in February 1958. In the Guangming daily newspaper the authors reported what the Chinese folk tales dating back to ancient times said about Bigfoot-like creatures living in the south of China. In several ancient Chinese books cited by the authors, we can find descriptions of "wild man". Although the authors have suggested it, there is nothing here that refers to the gorilla. The ancient texts describe clearly how the face and body has a shape similar to a person, although the mouth is wider. The body is also covered in dark hair. There are also some unexpected details. When they meet humans, they sometimes ignore them, but they can also smile and even laugh. They prefer to live in the river gorges in the mountains, where they sometimes turn over the stones to look for crayfish.

It should be noted that in a letter to the Commission for the Study of the Snowman Prof. Pei Wen-chung again stressed that Chinese scientists now have abundant information from various Chinese ancient texts about Bigfoot, "wild man" or "man-bear". The information is very similar, and resembles the modern descriptions. They are generally very realistic.

The Soviet specialist R.F. Its, who has studied the collections of ancient Chinese manuscripts and books in Leningrad, said that he had found a considerable number of references to the creatures, that can be compared to the "snowman".

In a very ancient Chinese book *Sou Shen Chi* (third century AD) the author Gan Bao recorded a long tradition of these beings: "In the kingdom of Shu (present-day Sichuan. – BP) in the high mountains of the south-west lives a creature that resembles a monkey. It is seven chi tall (1 chi = 0.32 m – BP). It can perform the same actions as people. It likes to chase people. It is called Jiagu or the clumsy one; it is also known under the name of Tsyueyuan" (Translation M.F. Sofronova).

The renowned Tibetologist Prof. Y.N. Roerich kindly informed us about the following extracts from old Tibetan books. From *Walking from Chaglotsavy Tibet to India* (the beginning of the thirteenth century): "On the way from Central Tibet to India, I had to continue the journey without companions; although the *e-gyo* (wild men) were plentiful, I saw no robbers." In *The Life of the Sixth Dalai Lama* (1683–1705) there is a story about how he once saw two humanoid creatures covered with thick hair. The Lama stayed strong but his companion fainted from fright. The Lama stood and watched the two beings. The creatures crossed the river, and collected armfuls of wood, and two medium-sized conifers, and then they crossed the river again. The Lama raised his companion, and they fled, followed by one of the man-like beings. The Lama and his companion managed to hide in a hermit's hut. The hermit told them that the beings were called "Mi-Tre".

Prof. Dr. Rinchen [12] of Ulan Bator (Mongolia) tells us: "In the collection of the late Dr. Zhamtsarano was a manuscript written by the Buryat monk who made a trip to Tibet and India in the ninth century. In this it was mentioned that he saw wild hairy men in the mountains of the Himalayas. This is the first written report about the Himalayan yeti. The monk also mentions monkeys and distinguishes them from the hairy people."

But back to the earlier handwritten testimony. In the ancient Indian treatise *Chzhyu-shi* ("Four Books"), which is a base of medical knowledge, there might actually be descriptions of the use of the internal organs of the "wild man". This is based solely on the fact that in later Tibetan and Mongolian medical books, the reader is referred to more detailed information on this issue in the Tibetan medical interpretation of the *Chzhyu-shi*, written in the tenth-eleventh centuries, and known as the *Lhen-Tab*. Unfortunately, the relevant information from the *Lhen-Tab* has not been found as yet. But it has been used by researchers studying the question of Bigfoot in the USSR and abroad. In the *Atlas of Natural Sciences*, published in the eighteenth century, we find images of different plants and animals used for the manufacture of different drugs, as well as drawings of medical instruments. All the plants and animals really exist in Central Asia. Next to each picture we find the Tibetan and Mongolian names corresponding to modern names. The images are stylised, but they are clearly taken from nature. Among several drawings of monkeys (rhesus and langur) is a drawing showing a two-legged humanoid standing on a rock, as well as one looking out from the other side of the rocks; maybe it's a male and female. The inscriptions state that this animal is called in Tibetan *mi-gyo* (wild man), in Mongolian *hsiung-guresu* (man-beast). Is it possible that this figure is drawn from life? It has appeared in books as woodcuts that historically precede the manuscript, so we can safely assume that the drawings in this book are not the originals, but have been redrawn from a more ancient manuscript. But all the pictures give a clear impression of the artist having a close knowledge of his subject.

We shall return to the question of these images later, but for now it is worth recalling the very instructive mistake made by European zoologists about the Asian tapir, simply because they ignored the image of it in ancient Chinese books.

In a more recent medical atlas from the nineteenth century the artist clearly no longer has any direct knowledge of the portrayed animals, but reflects the influence of various folk-tales. He has given them the faces of demons and turned them into more of an evil spirit than a wild humanoid animal. But there is a rather strange caption to the illustration: "The mi-gyo lives in the high mountains, it is a kind of bear with human form. It has extremely great physical strength. Follow the instructions for medical use of its meat and organs."

There is no actual contradiction between the statement that the animal is a "kind of bear" and an image not showing the slightest resemblance to a bear. Another example has been found by historians of ancient Chinese art: a pattern on a stone from the Han era clearly depicts a hairy humanoid standing on two legs with the caption "Bear Hunt". We also know of a series of other pictures showing various aspects of "bear hunting". In these the bear is portrayed as looking more like a human standing on two legs. In modern China the brown bear (*Ursus arctos*) is also known as the "man-bear", thus leaving room for a certain amount of confusion regarding the creatures described as "wild men". Ancient folklore can present a certain amount of difficulty for the zoological researcher.

It is possible to find a lot of information and news in ancient manuscripts and books, but also in oral traditions, and in the future a lot of this information will probably be used by researchers exploring the joint history of mankind and the animal referred to as Bigfoot. Among the Lepcha people in south-eastern Nepal Charles Stonor recorded an ancient legend passed down from generation to generation. "In our mountains we have a tall animal known to our ancestors as *thloh mungo*. In our language this means 'savage mountain'. Thloh mungo was so cunning and ferocious that he was considered a worthy opponent for anyone who would meet him. He could always outwit our hunters with their bows and arrows. They said that thloh mungo lives alone or with very few others like him. Sometimes he walked on the ground, and sometimes he would climb the trees. He can be found only in the highest mountains of the country. Although thloh mungo looks very much like a man, his body is covered with long, dark hair; he was much smarter than monkeys and much larger in size. Then more people were born, the forests were cut down, and the wilderness disappeared, and so did thloh mungo. But many say that this creature can still be found in the mountains of Nepal, far to the west, where the Sherpa tribe calls it Yeti." Stonor heard similar tales from other tribes in Nepal. "Many generations ago," probably 30, and maybe 70, no one knows, " the creature lived in the mountains, not far from the village. But when more people came all these strange animals left this place. People say he is now living somewhere in the far north, in the country of the Sherpas."

In the seventeenth century the German natural scientist Athanasius Kircher (1601–1680), one of the most prominent scientists and scholars of the time, wrote a number of works on physics and mathematics. In 1637 he was also one of the initiators of the first magnetic survey on a global scale. In Rome there is still a collection of objects of natural history and physical and mathematical tools belonging to Kircher. Apart from being a naturalist, he also studied history and archaeology in Italy, the history of Egypt, and taught oriental languages. In 1667 he published a book in Amsterdam about his travels in southern China. The book contains a comprehensive description of the country. Chapter VII, which is dedicated to the fauna of China, still contains information valuable today (it contains among others a description of animals now missing in these places, elephants and so on). There is even a section entitled: "Forest People". "In the mountains of the Fujian Province one can find the hairy animal-like Persian man (*Homo persus*); likewise it has been reported that in the provinces of Yunnan and Hunan one meets anthropomorphic creatures referred to as *fet*. They are taller than a man, with dark hands and hairy body, and they can run very fast. They practise cannibalism, and during a meeting with humans one gave off a sound like human laughter, and jumped onto a table. Pater Heinrich Roth told me that at a time when he was in Agra, the Mughal king was given a monster called 'forest man' (*Homo silvestris*)." Roth, who was a missionary, probably saw the creature in 1660 in the city of Agra in a procession of trained animals donated by the Mughal Aurangzeb.

Athanasius Kircher reached an interesting conclusion about the collected information: "I am well versed in the various stories, I think these beings must be some kind of wild huge monkeys, for their body is hairy, and they smile like a monkey with wrinkled narrow forehead, flat nose and bared teeth, and when these creatures are irritated or injured they make hissing sounds."

All the old evidence concerning the existence of a wild animal humanoid in the mountains of Asia is of course very fragmented. This part of the history of the Bigfoot has yet to be written in a systematic way. So far the information can only attest to the antiquity of human knowledge about this creature, but knowledge that is often woven into epics, folklore, and the lives of the saints.

And we haven't even tried to cover epics, sagas and ancient literature from other areas of Asia, where more information can be found, and no doubt will in the future. We have not even touched on the numerous cases in ancient Arabic and Persian literature. But that is a project for the future. They are generally known as *div* or *deva*, but it is difficult to ascertain the realistic elements, as deva are generally described as mythological creatures or spirits. But it is possible for researchers to detect traces of a biological reality in the various description of devas in the classics of Eastern literature.

In conclusion to this chapter we return to Europe and to an old book, which is located on the boundary between the pre-scientific and scientific periods. We are talking about the first edition of the *Systema Naturae* published in 1735 by the great Swedish scientist Carl Linnaeus.

Linnaeus was a physician and naturalist, and from 1739 he was president of the Swedish Academy of Sciences. Linnaeus established a scientific classification system of flora and fauna based on extensive work in botany and zoology. The classification system also included people, and described them as belonging to the mammalian order of primates. Linnaeus did not believe in any kind of evolution in the organic world. He believed that the number of species remained constant from the time of creation, and that the task of taxonomy was to reveal the established order in nature.

However, towards the end of his life, Linnaeus started to speculate on the mutability of nature, and that perhaps everything had branched off from an original species, and that new species were formed by crossing of others. This is a major departure from the metaphysical idea of the immutability of species that he had inherited from the medieval understanding of nature. It was a significant step forward in the development of science. The *Systema Naturae* was published in twelve editions in his lifetime, each with more and more species, and with a more and more advanced classification system. It was very important that Linnaeus had positioned man among creatures that were similar to him. It showed perhaps not the idea of man originating from apes through intermediary evolutionary forms, but rather that varieties and species already known could form a bridge between man and ape.

Thus Linnaeus described a special kind of man, *Homo ferus*, children raised by animals, with no human social skills but still with some human traits.

CHAPTER 3 • SUCCESSES AND FAILURES OF RUSSIAN SCIENTISTS

Science can only make new discoveries when it is ready. N.M. Przewalski made many trips in the nineteenth century. He was active for almost eighty years and made many great discoveries, including those in the field of zoology. At the time the biggest authorities in anthropology said that the Neanderthal skull belonged to a pathological modern man. The Javanese Pithecanthropus had not yet been found. One must also remember that the mountain gorilla was only discovered in 1901, and at that time there was no real theory about the origins of modern humans. Darwin's theory had scarcely begun to spread. It was at this time that Przewalski was on the brink of discovering Bigfoot, but did not actually do it.

In Przewalski's files there is lots of information from his travels in Central Asia, some of it concerning the “wild man”. He did not however consider it to be a reality, so he did not consider it worthwhile to search for it more thoroughly.

Most of the information comes from his first Central Asian journey in 1872, and especially the Chinese province of Gansu, where *hsiung-guresu* (man-beast) can be found living in the mountains of Nan Shan.

This is an excerpt from his book about the first journey – *Mongolia and the Tangut Country*:

> Even before our arrival in Gansu, we heard from the Mongols stories of an unusual beast, which is found in the provinces and called hsiung-guresu or “man-beast”. The storytellers assured us that this animal has a flat human-like face and walks on two legs. His body is covered with thick black hair, and the legs are armed with huge claws. It was extremely wild, and nobody dared to attack it, neither local residents nor hunters.
>
> The Tanguts in Gansu told us similar stories. They all claimed that the above-mentioned animal can be found in their mountains, although very rarely. When we asked if it was a bear, they said no. They knew the bear very well.
>
> When we were in the mountains of Gansu in the summer of 1872, we offered a reward of 5 lan to anybody that will show us the legendary hsiung-guresu. Nobody came to us with news. Only one Tangut said the hsiung-guresu live on Mount Gadzhur, where we went in early August. We found amazing animals, but despaired of ever seeing the hsiung-guresu. And then suddenly I heard that a small temple 15 miles from Chertynton had the skin of a hsiung-guresu. I went there with a gift for the abbot, and asked to see the rare skin. My request was fulfilled, but instead of a monster, I was shown a small bearskin stuffed

> with straw. All the stories of the hsiung-guresu were clearly a fable, and after I had assured the locals that this was the skin of a bear, they started saying that the hsiung-guresu did not show himself to people, but that hunters sometimes saw his footprints.
>
> The bear-skin was in a standing position, about 1.37 m high. The muzzle was elongated, the head and entire front part of the body an off-white colour. Back and legs were darker, almost black...Unfortunately, I could not make more detailed measurements and descriptions, as I did not want to arouse suspicion (?).
>
> In the spring of the following year we did see the same bear in the wild.
>
> According to the Mongols, these bears live in significant numbers in the Tibetan ranges. In the summer they wander out onto the plains, and can even get to the coast.

This story is rather surprising, and there is a clear contradiction between the statements of the residents that they are well aware of the bears in the area, where they can be found and what they do, and that they still used the skin of a bear as the skin of the hsiung-guresu. It seems to me that Przewalski simply took the first and easiest explanation he could find. He never thought that the animal may have had ritual significance, and that the abbot did not want him to see it, and instead showed him the bear and counted on the traveller not knowing anything of the various beasts. We became even more sceptical, when V.L. Bianchi found Przewalski's original diaries in the archives of the Russian Geographical Society. Based on these it seems clear that Przewalski greatly exaggerated his effort to verify the authenticity of the hsiung-guresu:

> August 1872: Moved onto the upper part of the North Ridge and stopped near the Gadzhur rocks, the highest and wildest part of the ridge. Our tent is now standing at a height of 12,000 feet. It is raw and rough, but it is a good place to go hunting and on excursions. According to the Tangut, the hsiung-guresu lives in these rocks.
>
> ... The ninth day of this damned rain ... another seven days, just sixteen days in a row with rain, it almost never ceases ... it is absolutely impossible either to hunt or go on trips. One is forced to sit idly by.
>
> This ridge (North Ridge Nan Shan Mountains. – BP) is much less rocky than the south. The Gadzhur mountain group is 25 miles to the east of the Chertynton shrine. At the western edge of the mountain among the enormous rocks lie two small lakes formed from mountain springs. The largest of these lakes is called Lake Demchuk ... it is considered holy, and each year the lamas of the Chertynton shrine conduct a ritual. (According to other reports, they

used the skins of Bigfoot at this annual worship – BP). In addition the Tanguts often come here to pray. Not so long ago, one lived in a cave here for seven years, the very Tangut that we had as guide into the Gadzhur mountains.

... Regarding the animals it can be said that there are few in the North and in the South ranges. The main reason for this is probably the many hunters ... Apart from marmots and deer, we found the same animals as in the Alashan mountains: deer, *cookie-Yaman*, musk deer and wolves. In addition, they say there are, although very rarely, bears, leopards and the fabulous hsiung-guresu ... Of small mammals I found only ferrets.

These diary entries are dated 2–16 August. On 19 August they are on the South Ridge. And here Przewalski makes it clear, that the attempt to survey the North Ridge actually failed. His observation about the hsiung-guresu is actually quite limited. He does mention the visit to the shrine, where the abbot might have tried to deceive him. The information is valuable though, even if his stories about the "man-beast" or "wild man" actually refer to the bear.

During his third journey Przewalski was again in Central Asia in 1879. He passed through Nan Shan and explored its northern and southern slopes. In his writings there is a description of an episode that until recently had not attracted much attention. During a hunt for wild yak and snowcocks, Yegorov, one of the cossacks, started following a bloody trail, lost his way, and wasn't found until five days later.

This episode can now be seen in an entirely new light as a result of a conversation between G.V. Parfyonov, who is an expert on Central Asian prehistory, and P.K. Kozlov [13] , another outstanding Russian traveller. This information was recorded in 1929. At the time Parfyonov was head of the Smolensk Memorial Room and Library. Kozlov had arrived in Smolensk to give several lectures about Przewalski's travels. "I, like others, noticed" said Parfyonov, "that whilst telling an episode about the disappearance of the Cossack Yegorov, Kozlov said there was something strange about the circumstances. Later, when I was alone with PK in the hotel room, I asked him what had happened to Yegorov. There was a lot more information in Przewalski's letter than in the official report, among it stories about wild people living in the reeds of the area. They did not have fire, they would catch fish with their bare hands and eat them raw. Kozlov told me that Yegorov apparently had met wild people, and we discussed whether they could be surviving prehistoric humans. The wild men that Yegorov had seen were similar to humans, but their bodies were covered with hairs. We also discussed their sounds."

It was hard to say why the story about Yegorov seeing wild humanoid creatures on the southern slopes of the Nan Shan mountains only came to us third-hand. One can assume that Yegorov had in fact seen the hsiung-guresu that Przewalski had heard of, but that the description of their appearance was somewhat exaggerated. Maybe they just showed up because of the intensive food stimulus. Yegorov had after all been hunting wild yaks.

During his fourth journey in 1883 Przewalski heard about "people" living in "a state of complete savagery" without fire or clothing, in the Lop Nor marshes in the Lower Tarim. We have not yet found Przewalski's original manuscript with the notes on this particular journey. According to Parfyonov, Przewalski had not made any specific data available in the printed text. This time he tied the information to traditional tales of runaway Buddhists hiding in the reeds from their Islamic pursuers. It seems that Przewalski for the third time failed to pursue information of the wild man.

Occasionally news from foreign countries appeared in the pages of Russian scientific and popular magazines. The journal *Nature and People*, published in 1899–1900, had a story about the alleged unknown human creatures brought from the Himalayas to Europe (all turned out to be wrong). In 1904 the magazine *Pictorial Review* reprinted a note from the *Daily Mail* with a correspondent's report from Lhasa, about stories told by Tibetans, that in some areas along the Brahmaputra River there are wild people who do not use clothes, have no articulate speech, no guns or weapons, and only use their teeth and nails in fights. But there were no reports from independent Russian scientists.

Of much more importance was a report in the press in 1899 from the prominent zoologist K.A. Satunin about the "Biaban-guls". They were a kind of human-animal living wild in the Talysh Mountains in South Caucasus. Satunin was the leading researcher of the fauna of the Caucasus, especially mammals. He described many new species. It would seem that stories like these could not attract the attention of an experienced field researcher, if they were not of interest. He cited not only stories heard from the local population, but also his own observations. The topics he covered were however not picked up by other zoologists working in the Caucasus. This thread is permanently broken.

In 1906 an event occurred, the value of which was only appreciated much later. From 1905–1907, a young scientist of Buryat nationality, B.B. Baradiyn, was sent on a trip through Mongolia to Tibet by the Russian Geographical Society. Later he told his friend, the Mongolian scholar T.Z. Zhamtsarano [14], that en route to the Labrang monastery in Tibet he had seen an unusual creature while his caravan was travelling across the Alashan Desert in April 1906. Unfortunately Baradiyn's description has not been preserved, so we can only make a picture of the creature based on the descriptions of the creature made by other persons and given to Professor Zhamtsarano.

In 1930, M.K. Rosenfeld, who was a correspondent of the newspaper *Komsomolskaya Pravda*, and had never heard of the yeti, put the following story by Professor Zhamtsarano in his book *By Car Through Mongolia*. "I could tell many stories about the wild *almas* known by the Mongolian nomads, but I am afraid that to you they would seem like legends, superstitions of ignorant people. But these stories are told by people who are convinced the *almas* really exists." The "wild people" are frequently mentioned in descriptions of travels in Central Asia. Many of these stories go back to antiquity, but some sources are very recent. In 1906 Professor Baradiyn was walking with his caravan in the Alashan desert. One evening just before sunset, when it was time to stop for the night, one of the men who was looking at the hills suddenly cried out in fear. The caravan stopped, and everyone saw a hairy man-like figure on a sandy hill. He was standing on the crest of the sand, lit by the setting sun. He was bending down, and his arms were long. For a moment the almas was looking at the people, but when it saw that the caravan had seen it, it fled for the hills. Baradiyn asked for some of the men to go after it, but none of them dared to do it.

Rosenfeld's record diverges to some extent from the record of Professor Zhamtsarano and the prominent Mongolian Professor Rinchen. "The late Professor Baradiyn was the only Russian scientist who actually met the almas during his travels. The caravan was travelling to the Buddhist monasteries in Tibet, and Gumbun Lavran and Professor Baradiyn went to visit the learned monk Sharab at the Urga monastery. He was famous for his extraordinary physical strength. While travelling through the Badain Jaran Desert in Inner Mongolia, Baradiyn was amazed by the appearance of a hairy man climbing a dune next to the caravan. Sharab explained that this was a wild man – an almas, jumped off the camel and rushed after it. However, with his heavy shoes and clothes, Sharab could not catch the almas, who quickly disappeared behind the dunes. Sharab did not dare hunt him alone, and the caravan continued on its way. For the Mongols, seeing an almas was as usual as seeing wild horses or yaks. Baradiyn later told his friend Professor Zhamtsarano about this, and I learned about it in 1927. In 1936 Professor Baradiyn contacted me in Leningrad and asked about whether his companion Sharab who tried to catch the almas was still alive. I asked an Urga lama if he knew this Sharab, and he told me recently that he had died as an almost 80 year old man, and he told me also that he was a man of great strength, and that he would fearlessly enter into combat with almas".

It is worth noting that in the cited article "Almas – Mongolian cousin of Bigfoot" Professor Rinchen discusses why there is no mention of this meeting in the published report of the journey of Professor Baradiyn. One reason was that he did not manage to photograph the almas, but there was also another reason. In an earlier report Professor Rinchen wrote: "Zhamtsarano said that Baradiyn wanted to include this incident in his report, but S.F. Oldenburg [15] dissuaded him, saying that no one would believe it, and that he should not mention the case to avoid embarrassment". A third reason that Baradiyn did not mention the almas in his printed report was that he called himself a Buddhist, and as such may have been told that, although the almas are not strictly speaking sacred, it was still not to be recommended to describe it in public.

Unfortunately, we do not have the archives of the late Prof. Zhamtsarano. However, according to Prof. Rinchen who is now continuing his research in this subject, Zhamtsarano did not keep extensive records, but preferred to make short notes on a map. "He had collected information from the nineteenth century to 1928 and marked out areas with suitable almas habitat. He had intended to make a pilgrimage to the these places which were labelled with the names of informants, mostly caravanners and wandering monks, who have heard or seen these strange creatures and their tracks… The Mongolian artist Soeltay, who was serving on the Committee of Sciences in Mongolia, has painted a picture of the Gobi almas based on information from Dr. Zhamtsarano. It has been marked - 'Almas based on descriptions by certain...'. Zhamtsarano was convinced of the existence of almas. He believed that the almas are dying out, and can now only be found in remote areas of the desert. Their extinction was in his eyes clearly demonstrated by his map, where he outlined the distribution of the almas based on information from each year... According to data collected over the years by the late Zhamtsarano from the 90s of the last century to the first quarter of the twentieth century the almas habitat was shrinking. It almost coincides with the habitat of Przewalski's horse and the wild camel in the south-western part of Mongolia... We planned that in 1929 a small group of two or three people should go to the possible almas habitat. Dr. Zhamtsarano was firmly convinced that the capture of almas would be of great scientific value. However, this expedition did not take place."

Dr. Zhamtsarano had several employees and researchers working on the study of the territory of the Mongolian People's Republic. First there was the other, now deceased, member of the Mongolian Committee of Science, Dordji Meiren. He has preserved for us, at least in summary form, the chronological and topographical information gathered by Zhamtsarano about the territory of Mongolia. "In the first part of the 1800s, there were almas living in many areas of southern and Inner Mongolia." Then, says Dordji Meiren, their number decreased, and points out a few places where they could still be found in the 1867–1927 period. In early 1927 there were so few sightings that he concludes the almas were extinct. Secondly, Dordji Meiren had also collected data about skins of almas stored as objects of worship in some monasteries of Mongolia – especially in Alashan in the Gobi. One of them is described as follows: "The hair on the skin was reddish, curly and longer than on a person. The almas skin was removed by a cut on the back, so that the chest and face are preserved. The face was hairless, but had eyebrows and long tousled hair on the scalp. The fingers and toes had nails, which were similar to humans." Thirdly, Dordji Meiren made the first attempt to prepare a Mongolian dictionary of scientific and biological descriptions of almas and "almaski" based on survey data. Thus, he wrote: "One of the animals of the Gobi is called the almaska, it is very similar in appearance to a human. The body is covered with sparse hair. The mammaries are long ... Another name is the almasok. So far none of the European researchers have seen or described almas".

I suppose Prof. Zhamtsarano wanted to send two of his young employees, Simukov and Rinchen, to the Gobi looking for almas.

A.D. Simukov had been a member of the Kozlov expedition. He later became an outstanding researcher of Mongolian subjects. We do not know how his interest in the Mongolian “wild man” started, and how much research he had done on this topic. It is possible that he had heard the same story about Yegorov's observations from Kozlov. Significantly, we only know that in Simukov’s travel records, relating to his later independent travel and collection of materials, there are a number of references to almas, including a description of their tracks. The full scientific value of Simukov’s material is not yet clear, but it is certain he attached great importance to the subject, since he intended to go on a special expedition to remote areas of the Gobi in search of almas.

Rinchen continues to collect survey data on almas in Mongolia, seeking to identify their geographical distribution and biological features. The “Information Materials” have published numerous data collected by Prof. Rinchen. Some of these will be presented below in appropriate chapters. Here it should be sufficient to quote some of the conclusions. “I collected information on almas in the Gobi until 1927 (and later in the 1950s – BP). They can be described as follows: almas are very similar to humans, but their body is covered with reddish-black hair, not thick – the skin shines through between the hairs. This never happens with wild animals in the desert. They are of the same size as the Mongols, but almas are stooped and walk around with half-bent knees. They have powerful jaws and a low brow. The eyebrow arches protrude compared to the Mongols. In females the breasts are long. When feeding, the females will throw their breasts over their shoulder when sitting on the ground, so that the young can feed while standing behind them. They do not know how to make fire.” There are also descriptions of their club-like feet and their fast run. Study of the almas and the Himalayan Bigfoot has been published in a recent paper by Professor Rinchen: “Almas – Mongolian cousin of Bigfoot.”

One could call the work of this entire group of scientists studying Mongolia a school but, according to Zhamtsarano, it would not be possible to study the question of the almas without at least a small link to a wider circle of Russian geographers and zoologists. Prof. Zhamtsarano had discussed the subject with the geographer G.E. Grumm-Grzhimailo [16]. There is indirect evidence of zoologist D.N. Kashkarova being interested in the subject. Finally K.V. Stanyukovich has information about a failed attempt by some travellers in Dzungaria or Kashgaria to capture a “wild man” alive, but it was killed after throwing some guides off their horses. There are also various types of information from the Pamirs: in the 1890s the zoologist Voskoboynikov collected information about the Rong-kul caves, and in 1909 the entomologist Jacobson heard stories about meetings with “wild men” in the Far East, but he did not attach any importance to the report, and did not try to find any details.

The zoologist V.A. Khakhlov managed to delve quite deeply into the study of the wild man in Central Asia. Khakhlov, who worked from 1907–1914 on the borders between Russia and China, collected reports of these creatures from Kazakhs, who wandered down to the southern spurs of the Tien Shan mountains, deep into Xinjiang;

but most importantly, he made the first serious comparative anatomical analysis of the reports.

At first glance it might seem that Khakhlov's reports on the Bigfoot had nothing to do with each other, and could not be used for generalisations and conclusions. But in fact his work did expand our knowledge of the problem, and raised the domestic interest in the hypothesis that great apes or relics were living in the mountains of Asia. Further information can be found in two documents found in the archives of the Academy of Sciences of the USSR, as well as in the testimony of the still living Khakhlov.

The first document is an application from Khakhlov to the Russian Academy of Sciences of Zaisan, dated 1 June 1914. In this Khakhlov presents his preliminary conclusions about the existence of *Homo erectus*-like wild creatures in Central Asia, comparing it to an animal looking like an "antediluvian man." Khakhlov asked the Academy of Sciences to send him the papers necessary for the organisation of an expedition to China to study this. He presented some examples of his records, and his main conclusion: "These stories, based on the words of eyewitnesses, are enough, I think, not to consider these kinds of stories mythological or simply fabrications. I think there is no question as to the existence of the *Primihomo asiaticus*, as one might call this creature."

This document will remain a milestone in the history of the study of the Bigfoot problem. It is important to pay attention to the title and the first words of Khakhlov's statement: "On the question of 'wild man'. By itself this question is not new. People, travellers and natives of Central Asia have speculated on it for centuries. The solution to this problem ... is very important for science". But why does Khakhlov believe that this is not a new question?

In 1907–1910 Khakhlov collected a number of data. The first came from the hunter-guides and from the Kazakhs who used to visit Dzungaria. In this area, they said, one could not only find wild horses (*at-gyik*) and wild camels (*thiers-gyik*), but also the wild man (*ksy-gyik*). Khakhlov wrote about this to his supervisors at Moscow University. "It is worth mentioning that in 1904, the then assistant professor P.P. Sushkin at Moscow University examined the Zaisan Valley and Tarbagatai. Since then, all of my work has taken place in collaboration with him. Sushkin maintained his correspondence with me, but after I arrived at Moscow University in 1909, he also established a personal contact, as I started to work under the direct leadership of his professor M.A. Menzbier [17]. Menzbier totally denied the existence of the wild man, whereas Sushkin reported that travellers in Central Asia had also heard about the existence of such entities. He was allegedly told about it by P.K. Kozlov. He advised me not to disregard any such information." We already know from above, that Kozlov and Sushkin had far more information from travellers in Central Asia than they had outlined in their reports. The late academician Sushkin, one of the foremost Russian zoologists and Darwinists, was one of the first researchers to have condensed this information. He later handed Khakhlov various documents found in the Archives of the Academy of Sciences of the USSR. One of these was a letter of 24 November 1914 from the University of Zaisan, where Sushkin at the time was professor at the university.

From this letter it was clear that it was not the first time somebody had discussed the problem and possible ways of practical research. Khakhlov does provide new environmental details, emphasises the difference between this type of information and mythology, and writes: “As you can see, the question has a different setting here, and it is based on pure facts, if you can believe just 1/1000 of what you have heard.” The author has no doubt that you can in fact trust a much larger percentage of the data. This exchange of letters caused a severe setback to Khakhlov's application to the Academy of Sciences. By order of the ethnographer V.V. Radlov it was filed, and Khakhlov was sent away. Radlov noted in the case file: “Notes have no scientific value”. Khakhlov and Sushkin now discussed another way of organizing the expedition – through the West Siberian Division of the Russian Geographical Society. As can be seen from the response, Sushkin and Khakhlov reported new information about “wild people” relating to the Kobdinskomu area in Mongolia – incidentally one of the last areas of Mongolia where, until very recently, there were reports of almas.*

In the following I will present the research findings of V.A. Khakhlov. For this purpose two papers will be used: "On the wild man in Central Asia", and "The Kazakh wild man". They were written in 1958–1959, based partly on field diaries, notes and drawings from the 1912–1915 period preserved in the personal archive of Khakhlov. These data have had very little influence on the contemporary literature.

For several years, the young researcher toured the areas of Zaisan and Tarbagatai in Dzungaria, where he carefully listened to and recorded everything he heard about the wild man. At first he of course believed that the wild man was a myth, but then “he heard of him being caught, saw the footprints he left in the sand, the smell he could spread, his yelling and resisting capture, and the fact that he lived for a while on a leash.” From then on he quickly abandoned the idea of the wild man being a myth. The local Kazakhs all had information about it, and they all turned to him as a knowledgable person and requested that he told them what kind of creature it was – man or animal, a beast looking like a man, or a man looking like a beast. “I personally did not see this thing. I could only judge it from the stories. I was perhaps at a greater loss than they because of the doubts about the existence of the wild man. They were convinced he is real and lives somewhere in the south, where there are also mysterious wild horses and wild camels.” Every time he tried to clarify the habits of the wild man, Khakhlov was told comparisons with the horse and the camel. Then he managed to find two witnesses among the Kazakhs, who had seen the wild man in

* Translator's note: There is a strange jump in the manuscript here – presumably a part of it is missing or has never been finished.

Central Asia where they had relatives. One of them had seen a female captive wild man on a leash every day for several months. The other captured one he had heard of was a male. Both of them saw the wild man in several different places, and did not know anything about each other. He also managed to find hunters and shepherds, who told him about other cases. It was a detailed and valuable survey with a compelling scientific conclusion.

One witness, Irene Kabyrga, was spending a night in the mountains a year before the survey with local herdsmen tending their horses. At dawn they noticed a person in the vicinity and, suspecting that it was a horse thief, they quickly jumped into the saddle, taking with them a long stick with hair loops at the end. The "man" ran awkwardly, so they managed to catch him with two loops. He shouted, or rather squealed ("like a rabbit"). The herdsmen told the visitor that this was a wild man, a harmless creature that nobody wanted to harm, so they should let him go. They released the wild man, but followed him until he disappeared, in a place with dry stalks and tall grass under overhanging rocks.

In a statement from 1914, Khakhlov described what was perhaps another case of a capture of a wild man. Some of the details are different. It is the story of one of twelve herdsmen who were camping in the wild. "I was so frightened I nearly fainted, when I suddenly saw a naked man." His comrades said that when he cried out, they could only comprehend that there was a man hiding in the reeds close to camp. The herdsmen jumped on their horses and stretched a rope all around the reeds, and began to drive the "man" through it. Suddenly someone pulled at the rope that a herdsman was holding and he nearly fell from his horse. The others came to the rescue and dragged a naked man overgrown with hair from the reeds. They threw a rope around his neck and tied his hands behind his back. Assuming he was a horse thief, they began to beat him with whips, demanding to know who he was. But the torture did not help, he just cried like a hare. In the morning an old man explained to the herdsmen that they had caught a wild man, and that he could not speak. The description of the wild man was very much the same as the description of the male captured above.

Another eyewitness found by Khakhlov saw the wild man in the area around the Manas river, or possibly on the shores of the lake. This time it was a female who lived tied to a chain in some small mill until finally being released into the wild. This creature rarely made any sound, but was mostly quiet and silent. Only when people approached would it grin and squeal. She slept in a very strange way, very much like "a camel" in the words of the witness, on her knees and elbows, forehead resting on the ground, and hands placed on the back of the head. (According to information young Tibetan children often sleep in this position.) This sleeping position explained why the skin on the elbows, knees and forehead was very rough, "like the sole of a camel". When offered food, the captive ate only raw meat, some vegetables and corn. She did not touch boiled meat, and only started to eat bread later. Sometimes she would grasp and eat insects. She drank water by leaning down and drinking like a horse, or by dipping one hand in the water and licking it. When she was released, she walked awkwardly with long arms swinging. Very quickly she fled into the reeds and disappeared.

Khakhlov also wrote about a retired Altai soldier, who in the 1860s had to go to Dzungaria, south of the lake. While he was out hunting for wild boar with an officer, they saw a strange woman covered with hair emerging from the reeds. She had a sloping forehead, deep sunken eyes, large jaws and a short neck. Her hands looked longer, and her feet were wide "as downtrodden shoes". When they caught her she squealed and struggled. As soon as she was released, she fled into the reeds with her legs wide apart "as if she had something heavy tied to each leg".

Khakhlov used a variety of methods to translate the various eyewitness descriptions from the Kazakh nomads and other people into the language of modern science. He realised that the storytellers could not have invented it all. Two people who were strangers to zoology and anthropology and had never met before could not invent stories with so many similarities. Khakhlov took photographs of gibbon, chimpanzee, gorilla and reconstructions of prehistoric man and gave these pictures to both witnesses and asked them to show which pictures were most like a wild man. Both pointed to the last, and said it was most like this, but different from it. Forty years later the same methodology was used by researchers studying the yeti in the Himalayas, and they got the same result.

From the witness descriptions, the zoologist must clarify all the differences in the body structure between this creature and modern man.

The head: The witnesses did not describe the forehead, eyes and nose, but said that what caught their eyes were the massive brow ridges and prominent cheeks. Instead of lips and chin they had massive jaws bulging forwards with a large mouth and powerful teeth similar to humans. Khakhlov attaches great importance to the psychological fact that these features attracted attention and impressed themselves so vividly in the minds of the witnesses.

When asked whether the wild man did not have any forehead, the eyewitnesses explained that they did, but that it was hard to discern, because it was very sloping and the eyebrows were covered with calloused skin with hairs growing from it. The witnesses also explained that the wild man has a thick neck, and prominent muscles near the nape of the neck.

The eyebrows arch, and the nose is small, but with large nostrils. The witnesses also described, that the jaw of the *ksy-gyik* was not like theirs, but looked like someone had cut off the lower jaw. They also stressed the protruding jaw and the big, very wide mouth. They compared the mouth with the mouth of a camel, adding that the mouth of the wild man was even wider. According to their description the lips of the wild man are very narrow, dark and almost invisible from the outside. They can only be seen when it bares its teeth. The teeth are much larger than human teeth, with front teeth angled “like a horse”. Males have large canines which can be seen protruding from under the top lip. The facial skin is naked and has a dark colour. There was not much information of the ears. They are quite large, with no lobes, and they seem sharp, perhaps because of the hairs growing on them. One witness clearly remembered the long black lashes on the eyes.

There is a significant lack of data regarding the nature of the hair on the head (as well as the eyebrows). From this he has concluded that it must be absent or significantly less on the wild man than on man. But one may ask, whether it is not in fact the fault of the interviewer. Did he perhaps forget to ask his informants about the hair?

On the basis of his enquiries Khakhlov made a sketch of the head of the wild man. Khakhlov stressed that it was only a draft, and did not have the full approval of the witnesses. They all said it resembled the wild man, but also differed from it. This article clearly demonstrates the objectivity of the researcher. But he never did find the explanation for the incompleteness of his success. I think the explanation is Khakhlov's desire to only look for differences between ksy-gyik and humans. And this led him to downplay or ignore the similarities.

When it comes to the upper and lower limbs of the creature, eyewitnesses noted that his hands and arms are long. They reach to the knees or even lower. A hunter who saw a wild man one morning on a rocky hillside said that he climbed the rocks very quickly “like a spider on a web.” This should perhaps express that the wild man threw his long arms forward, grabbing the ledges, and pulled himself up, with feet pushing the body along.

According to eyewitnesses, the hands and arms of the wild man are completely covered with hair, except for the bare calloused elbows and palms. One eyewitness compressed his hand laterally while bending the thumb into the palm of the hand, to show that the hand of the wild man was very long and narrow. When a wild man got hold of a rope, the witness said he did not grasp it as a person, but with all five fingers on the top like a hook. The first finger of the wild man is apparently not as opposed to the rest as in humans. But it is still stronger than the last three fingers, and that is important when climbing. The third and fourth finger Khakhlov assumes are of equal length. According to one eyewitness, the female would pick up small objects in a very peculiar way. The animal would press the object firmly into the middle phalange of the index finger with the thumb, and not use the ends as a human would. According to witnesses the fingernails of the ksy-gyik are long, narrow and convex.

Also according to witnesses the legs of the wild man were completely covered with hair, except for the bare soles and the callous formations on the knees. They noted that there were similarities between the feet of the wild man and humans, but also differences such as the widely spaced toes. The wild man will support himself on the ground on the bent fingers and the tip of the thumb. The length of the foot was equal to the length of one hand and the palm of another hand. The first toe on the foot is noticeably massive, short and pointing away to the side.

In other words, in the upper extremities the first digit is less opposable than in humans, but in the lower extremities more. Khakhlov believes that the three middle toes of the wild man play the main role in locomotion. The toenails were narrow, long and convex as on the hands, and this creates the impression of claws that are bent at the end.

According to the eyewitness accounts collected by Khakhlov, the general posture of the body of the wild man is not completely vertical. It seems to be tilted forward. At this point it is worth noting that Khakhlov seems to have exaggerated the morphological distance between the human-animal and man of his own accord. There is no eyewitness testimony to support this description. He goes on to describe the thick neck, which is tilted slightly forward so the head becomes drawn into the shoulder. It gives the impression of a man looking up, and also that he is ready to throw a punch at someone, just like a fighter ready to fight. The shoulders are shifted forward, creating a stoop, and this has been noted by several eyewitnesses. The wild man's hands are not hanging by his side, but a little in front. The animal is also generally rather short-legged and long-armed.

In general, I think the author has created an artificial ksy-gyik by underlining the characteristics that distinguish it from a human being, and not taking enough consideration of the similarities. However, Khakhlov's comparative anatomical study and reconstruction is a very important scientific work.

Khakhlov gathered a lot of valuable information about the ksy-gyik. Its colour is said to be like a dark reddish-brown camel with a touch of grey. The hair has structure like the hair of a camel foal. However, as all the sightings were done during the summer, this does not rule out a significant change in colour or structure of the hair in other seasons.

There was no specific information as to where this creature was seen. He could be seen near glaciers in the mountains, and in the desert or in reed beds close to water. The only important thing, according to Khakhlov, is that the area has to be deserted. When too many humans move into an area, the wild man will disappear. Ksy-gyik seems to live alone, or in more or less permanent pairs with or without young ones. The vast majority of the meetings with the wild man took place at night or at dusk.

There is no mention of a permanent home, but temporary lairs have been found in shallow caves under overhanging rocks, in hollows under thick bushes, in niches in clay , or in reed beds. In all cases the lairs emit an unpleasant smell.

In addition to plant food (roots, stems, berries), the wild man eats invertebrates, the young and eggs of birds, amphibians and reptiles, and what is probably the most important part of its diet – rodents. These can be found in the mountains and in the desert in enough numbers for him to survive. Khakhlov also has two stories about ksy -gyik eviscerating various animals, and leaving the insides before eating the animals – in these cases, chicks, doves and a gerbil, a small rodent.

Khakhlov also received some very interesting information saying that from the onset of winter, the wild man moved further south towards Tibet. In general he is also found more frequently in this area now. Professor Khakhlov considers it likely that his study of the Tien Shan wild man and the now widely discussed Himalayan yeti is in fact a study of the same animal.

"The more one is able to find out about this creature" – writes Khakhlov – "the clearer it becomes, that this is not a person, but a kind of humanoid creature. It could be a branch of the primates, which has been preserved from ancient times in Central Asia, and has adapted to a life in high mountains and deserts. This branch has developed human traits to some degree, which is why people who have seen the creature have taken him for a kind of man-beast."

To bring the project further Khakhlov planned to send a small expedition a thousand kilometres to the south, to Xinjiang, in the hope of catching and killing a peculiar two -legged animal whose existence he was in no doubt of, and bring at least the limbs and head back, by protecting them from degradation in formalin in a leather sack. One of the witnesses and a local hunter were hired to perform this difficult and unsafe journey.

But, as already noted, the proposals of the young zoologist from Zaisan had already been rejected by the Russian Academy of Sciences. Khakhlov has vividly described the pre-revolutionary conservatism of the Academy of Sciences. However, the First World War in 1915 interrupted all other attempts to mount an expedition. Yet it would be wrong to say that his work at the time did not yield any results.

We said that Khakhlov repeatedly informed Prof. P.P. Sushkin about their research problems with the wild man. In 1907 Sushkin sent him a long letter of 27 pages about the question of the origin of man, which showed he had taken a close interest in the subject.

In the records of the lectures of Sushkin about vertebrate zoology read in 1915–1919, Kharkiv and Simferopol have discovered a very peculiar interpretation of the evolution of primates, and the emergence of man in particular. All wood-living primates are deemed to be highly specialised deviations from the general line of primates leading to humans. The latter did not evolve from tree-climbing forms, but in the treeless mountains in cold areas. From there it spread to Europe, and then to the south.

The next attempt was made by Sushkin in 1922 in the article "The evolution of terrestrial vertebrates and the role of geological climate change." It repeats the thesis that the human ancestor was not a wood-living animal. Sushkin argues further that the structure of the legs of a man demonstrates that he was not adapted to climbing trees but climbing rocks, and from this the upright posture of man could evolve. So this was formed in the mountainous areas of Asia. Humans were not made extinct by later glaciations, as they had learned how to make fire. This also made it possible for them to spread out widely.

Five years later Sushkin published a new article: "High-altitude areas of the globe, and the question of the ancestral home of prehistoric man". This attempt to shed new light on the question of the transformation of the wild ape to man has left an important mark in the history of anthropological science. In this work Sushkin was forced to give up a large part of the previous two papers: indisputable anatomical and embryological data convinced him that man originated from a tree-climbing ancestor. But he insisted that man was formed in the rocky terrain and harsh climate of the mountainous areas of Asia. Geological upheavals made the cold climate forests disappear, so the Tertiary ancestor of man had to climb out of the trees and become a rock-climber, and later developed the ability to walk upright.

We already know that Sushkin had access to a large amount of empirical material about the possible existence of living relics of great apes in Asia. He acknowledged that he still needed confirmed skins or skeletons. Sushkin therefore did not give lectures or write papers based on this information. But he saw the similarities in the eyewitness descriptions, and their complete independence from each other, as an important argument in favour of their authenticity. And this again forced him to conclude that central Asia was the ancestral home of man.

Today it is easy to see the fundamental error that Sushkin made. He saw these relict Asian “wild people” as the remains of the direct ancestors of modern humans. But as anthropology has developed and more fossils have been discovered, it is clear they represent a “humanised” and endangered side branch. Sushkin's hypothesis was clearly artificial, and failed because he could not see these Asian “living fossils” as anything but a direct evolutionary phylogenetic link to modern man.

In a sense Sushkin's 1928 paper is the epilogue of the first stage of domestic studies of the yeti. The second stage took place in the 1920s and 1930s, consisting of monitoring and collecting information in the mountainous regions of the Pamir-Ala system in the Kopetdag area and various other places. However, the data collected in this period was still not properly focused. The third stage, which began in the 1950s, will be treated in Chapter 5.

Dr. I.K. Fortunatov, the noted expert on medicinal plants, has also shared interesting information with the Commission. Around 1935 Fortunatov lived in Karaganda, next to Lubsan Sharab Tepkin, where he was working on a Russian translation of Tibetan manuscripts, including descriptions of traditional medicinal practices. From discussions with him Fortunatov learned that from 1908–1912 Tepkin had gone through Buddhist spiritual training in Lhasa, and often visited "sacred" Buddhist places (lakes and mountains) to participate in funeral ceremonies. After returning to Russia Tepkin then went to Tibet in 1912–1913 and worked as a translator. On his return to St. Petersburg he was awarded a medal, and was later on, for a long time, associated with a group of Tibetologists. Tepkin was widely known during this time in Leningrad as a person very knowledgable in Buddhism, and widely respected by the socialist orientalists. It was through Tepkin that Fortunatov heard of "wild men" in Tibet.

When he first went to Tibet from the north with a caravan of merchants, he was in the northern foothills one night when a Tibetan driver woke him up so he could hear the cries of a "terrible lonely man". The shouts were sharp and prolonged, but incomprehensible. The driver told him the wild people could also be seen in the day. They are only found one or two together, they are overgrown with black or grey hair, they are tall and can run very fast. Sometimes they throw stones and shout very loudly and shrilly. They are only found among the rocks in the high mountains and desert regions. During one of his trips Tepkin was shown the tracks of the wild man on one or two occasions. When he was studying in Tibet, Tepkin had also seen and held several "wigs" (scalps?) of wild men. They were carried on the end of poles during certain ceremonies by people playing the role of wild men. They would also wear a mask so they would look like some type of funny and ugly creature. According to Tepkin, most Tibetans believe the wild men are a kind of failed cousins, that live a miserable existence. In the Buddhist doctrine, the wild man has no special place, but the people feel sorry for him. According to the Tibetan people, several of them have seen the wild men in recent decades, but 50 years ago it was different, then they attacked and...*

Much later, in 1949, Fortunatov met a second man, Valery Grachev, in Guryev. He also shared some information with him about the Tibetan wild man. In the last years before his death in 1960 he lived in Leningrad, and in conversations with V.L. Bianchi revealed some other information to the Commission for the Study of the Snowman. Valery Grachev, who was a Russian by nationality, had studied Buddhist teaching. He had received his higher education in St. Petersburg Oriental. Before World War I, he went to the East and spent many years in Central Asia – especially in Mongolia and Tibet to study the spread of Buddhism. He had lived in Buddhist monasteries to translate and understand various ancient books and manuscripts. When you are fluent in the Mongolian and Tibetan languages, it is easy to communicate with the clergy and the people, whom, he said, often did not know any European languages.

* *Translator's note: Here there is another jump in the text of the manuscript – presumably a section is missing.*

Valery Grachev had been interested in the question of Bigfoot even before he began his travels. This was mainly because of one of his teachers, the previously mentioned eminent geographer G.E. Grumm-Grzhimailo, told him the episode with the hairy wild men who threw stones at a convoy, as described in the book *In the Land of the Lamas. Travel to China and Tibet* by W.W. Rockhill we have mentioned before. [18]

So before leaving for Asia, Valery Grachev already had some information about the issue. Indeed he had reason to remember Rockhill in Buryatia, when the lama of the Agin Datsan (monastery) mentioned the *almas* as they are called in Mongolia. When he heard Grachev was going to Tibet with a caravan, the lama remarked: "Go. But do not fear the *almaasa*". Then he described the almaasa, saying it looked like a big man with hair everywhere except the face and palms of the hands. The lama said not to be afraid when they heard the cry of the almaasa, though it was very loud like a howl. According to the lama, the almaasa do not attack people. They are afraid of him.

In his years of travelling, Valery Grachev often happened to hear about wild humanoids, which in Tibet are called *e-gyo*. But, "true, I did not get to meet the mysterious beast. However, the lamas of monasteries where I lived and studied Buddhism and Tibetan medicine, always answered my question about these non-human animals, and talked about them as a reality. Many had seen them, and described them in detail with no mystical elements."

One winter Valery Grachev was living in a small Buddhist monastery on the northern slopes of the spurs of the Himalayas, near the north-eastern border of Nepal. From the old lama, with whom Grachev developed a very close brotherly relation, he once heard a story that took place in the convent fifteen years earlier (c. 1902–1903). For two nights, the garden beneath the walls of the monastery was raided by some animals eating the vegetables. "On the third night", the lama said, "I gathered everyone living in the monastery, and we took all the pots, copper plates and two chimney tubes and went outside. The animals were already in the garden. They ran about and rolled on the ground, all the while throwing vegetables, lumps of earth and other things into the air. I noticed that the animals had dark brown hair that was very long and thick. In the vegetable garden they ran on four legs, but sometimes they would get up and stand or walk like people. Then I realised that these beings are our cousins, and instructed everybody to bang all the plates and tubes together. The animals threw something at us and waved, and then they ran away. They have never been in our garden since then."

According to Grachev, pilgrims and wandering ascetics often meet wild men. They spend most of their life wandering, and tell stories of their meetings, not only in Tibet, but also in Bhutan and Nepal. I hear similar stories from monasteries in Central Tibet. They describe the wild man as a creature with a dark brown colour, round-shouldered, that feeds on insects, birds, and plant roots. They said he was living in the mountains, and that he was very strong, although about the size of an average person.

Some witnesses have found the corpses of what they call the *e-gyo*, drowned in the mountain rivers. We cannot say that they encounter the e-gyo at every step, but almost every monk would meet them at least once or twice during their lifetime. Typically, such a meeting would be when the monk was sitting quietly, without moving, absorbed in prayer.

"During one conversation, the monks remarked that if I do not get to see a live e-gyo, it was easy to see a mummy. It is kept in the Sakya monastery in southern Tibet at the Trom-chu river. The lamas there are said to have gathered a huge collection of mummified bodies of various animals, from flies to elephants, probably so that the monks could demonstrate all the stages of the faithful transformation of the spirit on earth. I have heard many stories of the mummies in Sakya, but I have not been able to go there myself." According to Fortunatov, Grachev only managed to see two or three of the wigs (scalps?) attributed to the e-gyo, but they were all very old, shabby and faded, and it was very hard to determine what they were.

"In conclusion," Grachev wrote, "in my view it would be wrong to disregard the belief of local people, that such a creature actually lives in their country." In the 1930s Grachev had repeatedly reported his data on the "wild man", but had only gotten negative reactions, and an influential Leningrad scientist had advised him not to do any more. He did submit his manuscripts, but at the moment they seem to be lost.

CHAPTER 4 • TRACKS AND TRACKING IN THE HIMALAYAS

We can now turn to the work and observations of foreign authors. This is an easy task, as a lot of reports have already been published. They can be found in West European and American literature. The best of these are undoubtedly the works of Heuvelmans and Sanderson. Other useful works are by Tilman [19], Morin [20], Izzard and Bordet. There is also a good overview in the book of Odette Tchernine covering quite a substantial amount of information. The little that is left out by the author is made up for in the afterword by Tilman. In Soviet literature, the main results of foreign studies of the yeti are best described in three articles by S.V. Obruchev.

The studies by foreign scientists can be divided in two stages: before World War II and after. We will also look at studies that extend geographically beyond the Himalayas to different areas of Central Asia, including Tibet, and to the more recent detailed work that mainly concerns parts of Nepal.

For some reason it has become a tradition to call Waddell's observation of tracks in 1887, published in 1898, the start of the European studies of Bigfoot. But that date has to be pushed back quite a bit. The start should probably be attributed to the diary by Fraser (1820) [21], the work of White (1835) [22], or of Hooker (1855) [23]. *

Prof. Y.N. Roerich directs our attention to the places where the author speaks of observations at an altitude of 4500–5000 m of two very large herds of monkeys, although he does not specify the species.

In a book by William W. Rockhill published in New York in 1891[24],he describes how he obtained information in Tibet in March 1889: "...I stayed in Lusara for about a week to make the final preparations for the journey to Lhasa. At the hotel I met an elderly lama, who was returning from Lhasa, to the Rav-Gombe monastery about five days journey south of Lusara. He described to me how he had joined a caravan in the wilderness in Northern Tibet. He also told me that several times during the journey the caravan met wild men with long matted hair that covered their bodies as clothing. They were naked, dumb creatures that looked like people, and they kept throwing stones at the caravan. They had no weapons that could cause any harm. I often heard stories about hairy wild men from the Tibetans I met in Peking, so the story of the old lama did not arouse my curiosity." [25] Rockhill was inclined to believe that all the stories of hairy savages were bears standing on their hind legs, but he forgot that bears could not throw stones at caravans. In relation to his stay in Mongolia, Rockhill says that here the residents saw numerous flocks of

* *Translator's note: Here there is once again a strange jump in the manuscript indicating a section missing or unfinished.*

yaks, wild asses, antelopes and *geresun bamburshe*. This term literally means “wild men”, and the locals described them as being covered with long hair, standing up straight and leaving tracks like humans. But they believed they cannot talk.

In October 1899 [26] an English traveller named Major L.A. Waddell who was studying the Himalayas in north-eastern Sikkim, on the border of Tibet, found large prints of bare feet in the Dongkha La Pass at an altitude of 5180 m. “A few prints of bare feet crossed our path and went towards the high peaks. The Tibetans attributed the tracks to a hairy wild man who they believe lives in the eternal snow." [27] But of course, Waddell was a firm believer in the superiority of the European mind over the “natives”, so he decided they were not capable of distinguishing the tracks of a bear, and thought no more about it.

In 1890 there was a British army report about colonial troops reporting strange wild and hairy apes having been killed by soldiers in the area around the construction of a telegraph line near the Jelep La Pass. They just dumped the bodies in the mountains. Apparently none of them was educated enough to examine the hairy ape closer. The British colonial troops had little use for research. This is supported by an experience reported in a British newspaper in 1905 by the traveller Hugh Knight: [28]

"I was coming back from Tibet, with another European and fifty porters. Close to Gangtok I fell behind to give my horse a chance to rest. Suddenly I heard a slight noise. Turning around, I saw a terrible creature about twenty paces away. It was a kind of man, almost naked, though it was November and it was very cold. He had yellow skin covered with hair and thick matted hair on his head; he had big hands and huge feet .... “ Knight recalls that when he arrived in the evening and told about his experience, he got the impression that the British officers on the border did not find these encounters especially interesting. So far we have not examined any English newspapers or military archives. For example we do not know from which issue of the newspaper *The Statesman* [29] Nicholas Roerich borrowed the following story of an English major. While travelling in the Himalayas the major rose from his camp at dawn and looked at a nearby cliff. He looked at the snow-capped mountains, and suddenly saw a tall naked man on a nearby inaccessible cliff. The stranger rushed down the steep cliffs in mighty leaps and disappeared. When the major questioned the locals and his servants he was told: “This is a snowman who guards the protected country."

If we return to the earlier reports we find that the first direct observation not made by an occasional traveller or scientist was made in 1906. Unfortunately the true records have been lost to science, and they are only known through later retellings. It was the English explorer, naturalist and geographer Henry Elwes who personally met a Bigfoot in Tibet. [30] Several British scientists, as well as relatives of Elwes, have seen the original manuscript describing the meeting and the details of the appearance of this creature and its tracks shortly before the First World War, but today these notes are lost. At a meeting of the Zoological Society of London, Elwes reported on the alleged existence of a great ape unknown to science in Sikkim. It was inhabiting the high mountains and only descended to the lower levels in extremely cold weather.

In the description of the 1921 Everest expedition led by Lt.-Col. Charles Howard-Bury, I found another description of mysterious footprints found during the ascent to the Lhakpa La Pass at an altitude of 6700 m. "The soft snow showed the tracks of rodents and foxes very well, and to our surprise also tracks that were very similar to bare human foot prints. The porters immediately said that these were the tracks of the *metoh-kangmi*." [31] *Kangmi* means snowman, *metoh* means nasty or disgusting. Since then, the expression Abominable Snowman has been part of Western European and American literature, although this term is not in widespread use in Tibet or Nepal. Here the population generally uses other terms, such as *e-kyo*, yeti and others. In April 1922 the next Everest expedition visited the Rongbuk Monastery in Tibet on the northern slopes of the Himalayas, and had a conversation with a senior lama. When asked about the metoh-kangmi the lama quietly said that five of these "wild men" (*mi -gyo*) lived in the mountains above Rongbuk.

By 1925, as cited in Chapter 1, N.A.Tombazi, who was part of a photogrammetric survey of the Kangchenjunga massif, was camping in Sikkim near the Zemu Pass. "My porters suddenly called me out of the tent. For the first few seconds the glare prevented me from seeing anything, but I quickly saw the object they were pointing out about 200–300 yards away at the bottom of the valley. It was a human-like figure holding itself perfectly straight, but occasionally bending over to dig up the roots of dwarf rhododendrons. On the snowy background it seemed dark, and was clearly not wearing any clothes. About a minute later the creature came to a dense thicket and disappeared. I studied the tracks left in the snow. It may sound like a superhuman, but the tracks only measure 6–7 inches." Tombazi gives a very detailed description of the outline of the tracks, but does not try to explain his observation or consider the possibility of it being Bigfoot.

In the 1930s there were a number of reports of tracks in the Himalayas. In the summer of 1931 an English pilot found footprints on a glacier near the source of the Ganges at an altitude of 4200 m. In 1936 the explorer Eric Shipton was returning from Everest when he found tracks on the Barchganga Glacier. When his porters saw the tracks they refused to go any further, saying that it was bad luck to meet this creature. A little earlier the ethnologist and botanist Ronald Kaulback had seen many footprints similar to those of bare human feet in the upper reaches of the Salween River at an altitude of 4800 m. Two of his four porters believed the tracks to be of a leopard, the two others that they were the tracks of a mountain man, a naked being that looked like a normal man, but with long hair on the shoulders, arms and head. Kaulback adds that there are no bears in these places, and zoologists have stated that leopards never descend steep slopes in groups of several individuals.

In the summer of 1937 Frank Smythe and his Sherpa porters found tracks in the Bhyundar Valley at 6000 m. The Sherpas immediately declared that this was the track of the *epka* or "wild man". Smythe very carefully measured and photographed the tracks. Three of the Sherpas were so sure of their identification that they asked to be able to write it down. "We (the names followed), accompanying Mr. Smythe saw tracks that we know are tracks of the epka or wild man. We often see tracks of bear, snow leopard and other animals, but we guarantee that these tracks do not belong to any of these animals." However, when Smythe handed over his material to the English zoologist Sir Julian Huxley, he, as well as the Belgian zoologist B. Heuvelmans, later said that they must have been bear tracks.

In 1937 one of the members of the expedition making topographic studies of the Karakorum found tracks that the Sherpas described as *de-ti*. A few days later they found other tracks that the Sherpas ascribed to a bear. In November 1937 John Hunt saw several hundred tracks at an altitude of 5800 m in north-eastern Sikkim during an ascent to the Zemu Pass. "Two animals had crossed the pass some time before me. The prints in the snow looked so much like human tracks that I took them for the tracks of members of the German expedition, who were in the area at the time. But after close inspection I saw that I had not been overtaken. In fact none of the members of the German expedition did come this way. After weighing all the evidence, I am not inclined to think that the footprints belonged to a man. There were a number of tracks, sometimes they would run parallel, and sometimes they would cross. If they were people they would have moved in single file. Later I asked the Sherpas about the yeti. Their stories were all identical, and gave me reason to assume that there is in fact a large animal, which is not a bear nor a langur monkey."

Also in 1937 tracks were discovered 800 km to the west near the source of the Ganges by Shipton and Tilman from the Karakorum expedition. Two Sherpas also saw marks on the ridge between the two glaciers. The tracks were three or four days old, and heavily modified by melting. They were 20 cm in diameter and were located 45 cm from one another. They thought it could have been footprints from a smaller species of Bigfoot. In July 1938 Tilman, who was returning from Everest, was walking with two Sherpas in north Sikkim on the way to the Zemu Glacier when he came across relatively fresh tracks, which he first took to be human tracks. But the tracks were on a very steep cliff. On his return to Darjeeling, Tilman began to make inquiries and found out that apart from him no one had visited the Zemu Pass at the time. Tilman only found out later that John Hunt had seen tracks in the same mountain pass a little earlier.

Unfortunately, we do not have any reproductions of the photos of the tracks taken by Smythe. But the arguments put forward by Heuvelmans and others, comparing the outline of tracks left by bears and man, are not convincing. What do we know about the peculiarities of locomotion in unknown species of bipedal primates, allowing us to to equate it with human locomotion? Identifying the tracks seen by Smythe with a bear would only be defensible if the track showed undeniable differences. The first toe of the human foot is much more massive compared to the other toes on a human foot than it is on a bear's foot. This shows that humans are evolved from tree-climbing primates. At a meeting of the Commission for the Study of the Question of the Snowman Prof. A.N. Formozov demonstrated a series of tracks of different types of bears. There is no doubt that the shape and size of the first toe is of high value when identifying tracks of Bigfoot. When this is present, this cannot possibly be a bear. Therefore we were surprised to see a schematic illustration in the book of Heuvelmans, of a track of Smythe's "bear" with a clearly enlarged first toe.

Scientific and mountaineering expeditions to the Himalayas had stopped during World War II, and only began again on a small scale in 1949–1950. In the summer of 1951 a large British expedition went to the Himalayas with Shipton, William Murray, Michael Ward and Edmund Hillary taking part. On their way to the highlands of Nepal, the expedition visited the Tengboche monastery in the Khumbu Region and heard additional information about the "snowman", although the expedition was not planning to study this subject. A most interesting sighting had occurred in 1949. During a traditional holiday there were many Sherpas staying in the monastery. Suddenly they saw a yeti in a thicket of trees only 25 feet away. One of those present, Sherpa Tenzing, was later subjected to a detailed cross-examination, which left Shipton in no doubt of the sincerity of his testimony. Tenzing was sure it was neither a bear nor a monkey, both of which he knew well. It was animal-like and about the height of an average person (1.65 m). It was standing on two legs, had no tail and a pointy head. The whole body was covered in reddish-brown hair except the face. According to Tenzing, the yeti mostly moves about on two legs, but when in a hurry uses all four. Tenzing and his companions beat on drums and made a terrible noise to scare the yeti and drive it away. But it was not very frightened, and only disappeared after a while.[32] It is noteworthy that two years later in the same monastery John Hunt wrote about another monk who had seen a yeti approach the monastery walls.

One of the more interesting observations was made by Shipton: "One evening we were crossing the Menlung Glacier at an altitude of approximately 19,000 feet when we saw a line of footprints in the snow. We followed the trail for about a mile until it was no longer safe to walk...I have in the past found a lot of interesting prints and have tried to follow them, but have always lost them. The tracks seemed very fresh, probably no more than 24 hours old. When Murray and Bourdillon passed us a few days later, they were a bit more blurred." Shipton managed to take some extremely valuable photos. "The tracks had an extremely fine quality, because they were imprinted on a thin crystalline layer of snow on top of solid ice. They must have been made just a few hours before we found them. Each track was about 75 cm from each other. About 550 m above the forest we came to a place where two beings may have been walking together. At some places the tracks were close together, but later on they crossed again. At one point one of the creatures had jumped a crack with a width of about 90 cm. On the other side there was a clear handprint where the animal had landed in the snow. My Sherpa had no doubts about the tracks. One should remember that the Sherpas know the tracks of bears and monkeys well. They unhesitatingly called the being *de-ti*."

Shipton was lucky enough to photograph the tracks when they were very fresh, and had not been distorted by melting snow. Furthermore they were found on a very shallow layer of snow, whereas deep snow usually makes it impossible to see the exact shape of the foot. Shipton also photographed a long chain of tracks, showing clearly how the animal had moved bipedally. Shipton's pictures originally appeared in *The Times*, but have since then been reproduced countless times. Murray writes about the same tracks – first about how they found Shipton and Ward – later about how they crossed the Menlung Glacier and followed the tracks for about 2 miles.[33]

1952 was marked by two Swiss research expeditions. The first of them, led by Dr. Wyss-Dunant, was particularly important. They were in the region of the Khumbu Glacier in April where they studied several long footprints of several individual Bigfoot. They were of different sizes. The biggest specimen had a plantar (sole of the foot) length of 25–30 cm and a width of 12–14 cm. The next one did not reach a length of 20 cm. The depth of the tracks in the snow pointed to a large animal weighing 80–100 kg. All five toes were visible on the tracks, with the first turned somewhat inwards. Some of the tracks had imprints of nails, and a clearly marked heel. Analysis of the tracks showed that it used all fours when jumping. Wyss-Dunant suggests that the yeti does not live alone, but in families, as seen by the tracks of various sizes. He suggests the smaller individuals are the young ones. "I could not find any food residue or excrement. This supports the hypothesis that these animals are just passing through these places, and that they do not often climb to these heights. If the yeti lived and hunted in the area all year round, we would find the den or at least some temporary shelter. I am inclined to think that it only climbs the mountain passes when travelling from one valley to the next in search of food. It avoids places inhabited by people, but is apparently not afraid of mountain areas with large predators like leopard."

Simple logic demands that a large mammal cannot be a permanent resident of the eternal snow. There is nothing to eat. There is no doubt that it only uses the glaciers and snowfield when moving from one place to another, such as when migrating or avoiding danger. But footprints in the snow are a dead end. No matter how much one studies the rare and random tracks in the snow, it does not lead to the discovery of the animal or knowledge of its biology.

During the second Swiss expedition in 1952, one of the porters, Tenzing Mink, seems to have met a living yeti. The leader of the expedition said he was personally convinced that there is no reason not to believe the story.

1953, which brought the long-awaited victory over the world's highest peak, Everest, also contributed to the study of the Bigfoot problem. First of all Tenzing Norgay, one of the conquerors of Everest, reported on the de-ti. John Hunt, the leader of the Everest expedition, also recorded a story by the assistant abbot of the Tengboche Monastery. [34] This is the first time researchers turned their attention to the Sherpas and their knowledge of Bigfoot. Secondly, the Indian group of climbers led by Rusi Gandhy, and then the party led by the British Everest expedition doctor Charles Evans, visited the Nepalese mountain monastery in Pangboche to see an ancient scalp, allegedly from the head of a yeti shot a long time ago, and now considered a relic and used in religious ceremonies. He managed to take numerous photographs, and cut a few hairs for laboratory study. The annals of the study of Bigfoot would be incomplete, if we did not mention the results of a three year (1950–1953) ethnographic study by the Czech Professor Rene von Nebesky-Wojkowitz in Tibet and Sikkim. He collected information from the population with the material brought back by the English expedition.

1954 was marked by a sensational expedition to the Himalayas specifically to search for the Bigfoot. It was sent at the expense of the British newspaper the *Daily Mail*. It was led by Ralph Izzard, a representative of the newspaper, and had several knowledgeable naturalists: the British Charles Stonor and William Edgar, an Indian zoologist called Biswas and an American, Gerald Russell. The expedition was well equipped and widely covered in the press. In a sense, the Izzard expedition was in fact the culmination of the previous cycle of exploration and research. The results of the expedition, which spent the period from January to May 1954 in the high mountains of Nepal, were set out in two books, one by Ralph Izzard [35] and one by Charles Stonor. These books are widely known, and there is a Russian translation. This eliminates the need to set out in detail the fascinating history and interesting results of the expedition. We will only look at its most important conclusions.

First of all, Stonor managed to gather a significant amount of data from Sherpas from at least ten high-altitude villages. Combining his interest in ethnography, anthropology and zoology, Stonor managed to enter the world of the Sherpa's ideas of the Bigfoot deeper than anyone before him. He started out believing that these ideas should be attributed to the field of myths and legends, but came to believe in the actual existence of this animal. Stonor ends his book with the words: "The appearance of the yeti, as described by the Sherpas and Tibetans, coincides to a surprising degree with the appearance and way of life of some species of bipedal primates who lived in ancient time as recreated by paleontologists. For some of us, it is evident that the Sherpas would not just invent an animal so similar to those who existed in antiquity, especially because the Sherpas could not have learnt about these animals from other sources. The Sherpas describe the yeti and his habitat in a very realistic and detailed way. We have a lot of circumstantial evidence of their existence. Summing up, I can only assume the point of view of the Sherpas. They do not need to invent a new breed of mammal if they are not interested in it, and prefer not to meet with it. Their idea of the yeti is not based on mythological associations, but on the existence of a true beast of flesh and blood."

But Stonor's success had a down side to it. He was convinced that the local population is opposed to discussing the yeti and is extremely wary of questioning. However, Stonor managed to proceed by giving broad assurances that his intention was never to go further than to photograph the yeti. Thus the local population was not only a mediator between the researcher and the animal, but also a barrier. Sherpas are Buddhists and will never kill wild animals. They do not want foreigners to be a threat to this animal, which occupies a special position. If we want them to share their information, it is important not to overstep this boundary.

The second line of study of the 1954 expedition was to try to get to study the "scalps" and other physical remains kept as ritual objects in various monasteries in the Nepal mountains. These have been mentioned before. Photographs of the Pangboche scalp were, as I have said, published in the previous book in 1953. Izzard and Stonor have described their futile attempts to break through the superstition and get proper information about the precious objects. They managed to take new photographs and hair samples, which will be discussed in one of the following chapters. They had to realise that Buddhist superstition was a nearly insurmountable obstacle to scientific research.

The third part of the expedition was the attempt to find tracks and then track the Bigfoot. When it comes to finding new tracks the expedition was very lucky. They have done many observations of how the yeti moving in snowdrifts resembles swimming, how it would glide on its behind down snow slopes to bypass human habitations. In the Dudh Kosi valley Izzard and Russell walked for two consecutive days on the trail of two yeti; in total more than 13 km. The snow read like a two-day story of the lives of these two creatures. But catching up with the yeti in the snow was a hopeless undertaking. They did not dwell here or look for food. The expedition concluded that the yeti does not inhabit the zone of eternal snow and ice. They live below on rocky slopes with shrubs where they can find plenty of food. But here they leave no tracks, so it is impossible to find them.

The main conclusion of the expedition was in fact rather pessimistic. "When we left England, we thought we could search for them by finding tracks in the snow and following them until we encountered the animals. In fact the Sherpas' view, that the yeti wanders at random over a vast area studded with rocks, above the forest and below the eternal snow, is the correct one." Izzard also wrote a warning to future expeditions: "In our opinion, the yeti is likely to be found by accident, when someone stumbles upon him, and not as a result of organised searches."

This conclusion essentially was an admission of complete helplessness. But there could be no success, as they had ignored one of the most essential parts of zoological work of this kind. It is impossible to make a search without at least a theoretical hypothesis about the biology of the creature one is looking for. Searching blindly is of no use. Surprisingly Izzard never talked about using bait either. The 1954 expedition clearly showed why there was a crisis in the study of Bigfoot.

The whole idea of believing in luck and fortune, and forgetting the importance of preliminary generalisation and establishing hypotheses about the lifestyle and habits of the target animal, was manifested in the termination of the European and American researchers gathering or at least publishing collected information. A continuous and patient collection and analysing of data would at least lead to a gradual refinement of the map of distribution and the probability of migration, the daily cycle of activity, number and location of younger animals and so on. Unfortunately, the researchers decided not to treat this in a serious manner.

In 1955 two members of an Royal Air Force expedition photographed a number of tracks in the Kulti Valley at an altitude of 12,000 feet. According to the report, there were a number of tracks and evidence of bipedal movement. Five toes could be discerned on the feet, and the tracks measured 29 x 14 cm. Unfortunately they were sunk in very deep (27 cm) snow.

In May 1955, a French geological expedition to Mt. Makalu found tracks of Bigfoot around the Barun Pass. The tracks were studied by geology professor Pierre Bordet, who had not intended to address the Bigfoot problem at all. Unfortunately the tracks were old and badly eroded. On his way down with his Sherpa Ang Bao and several porters, Bordet met the postman, who had an escort with him, as he was afraid to pass this section of road because of the yeti. The next day, the French found a set of fresh tracks crossing the tracks left by the postman, so they must have been made the night before, or possibly even in the morning on the day of observation, meaning the sun had not had enough time to deform them. After a short inspection Ang Bao said: "this - yeti, go, you cannot stay here," and very quickly he and the porters left the area. More details could be found in a detailed paper published by Professor Bordet in the *Bulletin of the Museum of Natural History* in Paris:

> I'm going on the trail toward the east, traversing a fairly steep (20°–30°) slope. The imprints of both feet are clearly visible, pressed into 10–15 cm of snow. The animal slipped a few times, as indeed did I, although I was armed with an ice axe. At one point, it had reached the upper edge of a small rock wall with a height of 1.5 m; it just jumped off and continued on its way; no trace of the front paws. I'm forced to descend via a crack located to the right. After 200 m the tracks turn sharply to the left and go up. They disappear above 200 m, when going over the rocks, up the increasingly steeper slope... We're going on, notice a new track from the other side of the lake. Again, I leave the porters and walk towards the tracks. The footprints reduced to a sequence of depressions embedded in the snow. There's nothing to see... So, I followed the tracks of the others for over one kilometre, and saw about 3000 prints. They all look the same. It is a deep imprint of a foot, which has some similarities with the human foot. The sole of the foot is roughly elliptical and has a rounded back part. In front of the sole are four prints (instead of five) of roughly circular toes. The first toe is thicker than the others, and set slightly back. The other three are

placed on the front contour of the sole and very close to it. They are much thicker than human toes. Claw marks are not visible. On the best tracks, small snow walls could still be seen between the imprints of the toes. Print length about 20 cm. The step length is about 50 cm, somewhat less than the length of my steps on slippery ground. However, gait seemed uncertain, the animal is clearly in no hurry, and not scared as seen by the fact that it crossed the tracks of people. No trace of tail could be seen. The second set of tracks I saw belonged to an animal going down to the lake, apparently to drink. The prints were placed on a strict line and the step length was much larger, they must have reached a metre. This trail was lost in parts, where the snow melted. There is no doubt that all of these tracks belong to bipedal animals that, even in difficult conditions, do not feel the need to use their forelimbs... The yeti prints E. Shipton photographed in 1951 are undeniably analogous with those that I have seen: the same overall shape, the same number of toes - four, not five.[36] The next, photographed by Shipton is however different, as it is much larger in size and the position of the first toe is further back. One might think, therefore, of variations, depending on the growth of the animal or perhaps on sexual dimorphism.

Bordet has also collected information among the Sherpas, and he writes that they consider it as just an animal living in their country. They know where it lives, its habits and its different cries. Bordet concludes: "The yeti is apparently an unknown animal of which a fair number must exist. What they represent at this level of information is too early to judge, but simply denying the existence of the Bigfoot is not logical."

In 1956, Norman Hardie was part of a British expedition in Nepal. In the published report [37] there is only a little material related to Bigfoot. The author himself did not see tracks, but describes the sightings of others, who saw tracks about 23 cm long with a step length of about 90 cm. The observer could clearly distinguish them from the track of a bear, and for various reasons they could not be tracks of a langur. Norman Hardie also reports information of another "scalp" allegedly kept at the Rongbuk Monastery in Tibet on the northern slopes of the Himalayas.

In 1956 two persons were destined to have their names very closely linked to the yeti and Bigfoot research for the next four years. One was the Texan businessman Tom Slick, who travelled to the area and gathered evidence from caravans and local people who had seen the yeti. The second one was the Irishman Peter Byrne, a hunter and traveller, who knows Nepal well, and can perhaps talk to the population a bit easier. I have met Tom Slick and Peter Byrne, and I am convinced of the value of their data.

In 1957 they went together on an expedition funded by Tom Slick. Once again they found and photographed footprints of the yeti. "We walked in the footsteps of the creature in three places," wrote Tom Slick in a report, "for as long as they were not lost on rocky soil, or we were forced to stop the search at nightfall. Walking along one track, we found five-fingered hand tracks like from a human, at a place where the yeti slipped and slid down a steep slope. We found and collected fur clinging to the rough bark of a rhododendron. We found the place where the creature was leaning on one arm or leg to keep his balance in the soft snow, while it tore off moss-covered tree-trunks. There seems to be two types of track, one large – 30 cm long, and the smaller no more than 25 cm, but wider than the first, with a distinctive big toe branching off in a Y-shape. The last tracks," says Tom Slick "have an extraordinary resemblance to the fossil traces found six years ago in a cave near Turin, Italy, which are considered traces of Neanderthal man." [38] They found 15 witnesses, who claimed that the yeti is 2 – 2.4 m tall, but that there is another similar being 1.5 – 1.8 m tall called *mi-te*. The Sherpas were also shown a selection of photographs and pointed out the similarities between what they had seen and pictures of gorillas and prehistoric man. Some Sherpas even described the call of the yeti.

In 1958 Tom Slick financed another expedition which included members of the 1954 expedition: the naturalist Gerald Russell, brothers Peter and Brian Byrne and others. One group did stationary observations in the Barun Valley. According to Peter Byrne, a yeti came close to their tent twice during the night, but he did not have time to see it in the light of their lanterns. The second time they found 25 cm long tracks had passed very close by their tents, continued on, and disappeared on a bare rocky slope. They also found excellent thumbprints close to the tents. Russell had been in the Choyang valley with two Sherpas and had seen a yeti several times at night, once in the light of a torch, the second time in the light of a lantern. In the morning they had photographed tracks of a large animal followed by a second smaller one. Members of the 1958 expedition tried to invent new search methods, but as they had not taken ideas of the animal's biology into account, they invariably failed. First they wanted to use specially trained dogs used for bear-hunting brought over from the United States, but they were unprepared and completely useless in rocky terrain. Secondly they tried using bait, large frogs that the yeti would apparently eat, catching them in the water or under boulders. They tied a nylon fishing line to a frog, so that if it was grasped by a yeti, a gun with a paralysing agent would be fired. The experiment was a failure, probably because frogs are not a regular food item for Bigfoot.

The 1958 expedition also collected survey data, but almost nothing has been published. They made a series of observations of new tracks, but only confirmed that they could be separated into two animals of slightly different size and shape. One of the most important results was a fairly good plaster cast of one of the two types. But the real scientific achievement of the expedition was something else. In 1954 the English Tibetologist Professor Snellgrove learned that in addition to the scalp in Pangboche monastery, the monks also kept a mummified hand, but that it was wrapped in several layers of material and tied with a cord, and that to remove the bindings would be to desecrate the sacred relic. Thanks to Peter Byrne's knowledge of the Nepali language, he managed to have the hand released from the bandages, and he was allowed to photograph it several times. Moreover it became possible to remove the dried fabric from the back of the hand. Photos of the skeletonised hand were taken, and several particles of dried soft tissue were removed and taken to the United States and subjected to rigorous laboratory, anatomical and morphological analysis.

In 1959 two Japanese expeditions, one of which was led by Professor Teizo Ogawa, found new tracks and made several casts, and Prof. Ogawa was also allowed to photograph the mummified hand in Pangboche.

Using the methodology of the 1954–1959 expeditions it is clearly not possible to hunt down and maybe shoot or catch a Bigfoot. Several of the researchers lost interest, and in the following years no real results were reported.

This is not just a matter of lack of finances. The regular appearances of expeditions in the same valleys in Nepal has had other implications. In the traditional areas fewer tracks are being found. Possibly because the yeti has moved away from the area and people who were interested in them, but who did not take into account that their presence could alter the behaviour of the animals. There has also been a growing opposition within the Nepalese spiritual and secular authorities towards further searches for Bigfoot. The government of Nepal has announced that anyone who wishes to look for Bigfoot will be charged 5000 rupees. It is also strictly forbidden to kill a "snowman", unless it attacks. Residents of Nepal are no longer allowed to give information to foreigners without government permission. Any yeti alive or dead, as well as photographs, has been declared state property, and cannot be taken out of Nepal.

So any research in Nepal has effectively come to an end. And yet in September 1960, a new expedition in search of the yeti went to Nepal led by the famous mountain climber, the conqueror of Everest, Sir Edmund Hillary. Unfortunately, the world press started writing about the expedition. There was much speculation and suspicion. Many members of the expedition turned out to be missile experts surveying the borders of China.[39] Was the Bigfoot expedition just an excuse for their work? It was also quite surprising that Sir Edmund did not bring any experts on the yeti or any researcher who had taken part in previous expeditions. And yet there were more than 600 people in the expedition. There were a few zoologists. Marlin Perkins of the Chicago Zoo, who was not especially qualified, and a certain Dr. Lawrence Swan.[40] The expedition had a special reporter called Desmond Doig. He was familiar with the Nepali language, but had no time to read much about past expeditions. The expedition was financed by the Chicago Books Publishing Company, and they had set Hillary a simple task. He had to return either with proof of the existence of Bigfoot, or with a rebuttal.

After spending a short time in the mountains, not finding anything new, Hillary went to the Pangboche monastery and announced he was taking the famous scalp attributed to the yeti for six weeks to America and Europe for studies, and that this would decide the matter. [41] It now appears that Sir Edmund knew more than a year earlier that the scalp would be fake, and that it was even made to his order.

According to Mr. Sanderson, "the whole thing turned out to be a farce, if not fraud". Two papers written by Bernard Heuvelmans and Ivan Sanderson, two major western authorities on Bigfoot, have shown how much damage this has done to science.

CHAPTER 5 • THE ASIAN PROBLEM

Soviet scientists were not involved in the studies carried out in the Himalayas and Karakorum. But these studies drew the attention of the world to the question of the Himalayan Bigfoot. In early 1955 corresponding Academy member Alexandrov made a presentation and urged the Academy not to dismiss the Himalayan results, although there were a lot of meaningless writings in European and American newspapers. A number of our reputable scientists expressed their views for and against the existence of the Bigfoot in the press. The most comprehensive scientific information on the results of the Anglo-American studies in the Himalayas was given in two articles by USSR Academy of Sciences corresponding member S.V. Obruchev (Proceedings of the All-Union Geographical Society, 1955 and 1957).

We need to consider the hypothesis in a zoogeographical light. If we assume the existence of a previously unknown species, we must try to determine the geographical area of its distribution. What are the geographical limits of the observations? If it is possible to establish the geographical distribution, it must belong to specific areas (habitat) occupied by specific associations of plants and animals. This in turn enables us to speculate on the ecology of the species and its relations to the environment.

The basis of modern advanced biological science is the study of the connections between an organism and its environment. Once zoologists studied every species of animal by itself. Skins or skeletons of dead animals would be brought to a zoologist to be described and systematised. Modern zoologists see the morphology and physiology as one side, and its ecology ("ecological niche") as the other side of an organism. Both sides form an inseparable entity. This is how we should conduct the scientific approach to the mystery of the Bigfoot. If we can make conclusions about the environment of the creature, it can lead to recommendations of new methods of luring and surveillance, and possibly break the deadlock that has been in existence for so many years.

Two questions relating to the Bigfoot must be considered. First whether the research and observations made and collected in the 1950s from the mountainous regions of the Himalayas and Karakorum are of any value. Secondly whether the information from foreign travellers and researchers corresponds with the information previously accumulated by our domestic scientists and geographers regarding the distribution of the Bigfoot.

It turns out that it is only possible to see a clear boundary to the south of the studied areas. The Himalayas form a kind of a steep wall in the northern part of the Indian subcontinent. They form a border with the huge Indo-Gangetic lowlands. Here, within a distance of only 100 km there is a sharp difference in climate, vegetation zones and human population. At the foot of the mountains there are tropical jungles with their characteristic fauna, and at the top, starting from 3000–4000 m there is an alpine vegetation zone, which is rarely visited. When you go higher, you come to the zone of the eternal snow and ice. However, to the east, north and west there is no clear border area.

When the French geologist and mountaineer Professor Bordet summarised the material about the Bigfoot until 1956, he wrote: "In the classical literature, about fifteen Europeans have seen tracks of the *de-ti*. The outer limits of the area where they have been seen are: in the west, the Biafo Glacier in the Karakorum; in the east, the Zemu Pass in Sikkim. Further to the east, there is not enough information." Travel in Bhutan is very difficult, and we do not have any promising data from this area as far as we know. But there is definitely more information as we go east from Nepal towards Sikkim. Izzard also drew attention to the fact that in the past Sikkim had just as many sightings as the whole of Nepal. The Indian zoologist Professor Biswas heard many stories about the yeti in Sikkim, but did not formulate any opinion about the creature. The eastern boundary of the creature's distribution seems to go far to the east of the area which has already been studied. Actually we already have some fragmentary information not only from Bhutan, but also from mountain areas of Assam, Burma, Cambodia and Vietnam.

As for the northern border, there is a general consensus that the Bigfoot dwells exclusively on the southern slopes of the Himalayas, and only ventures onto the northern slopes more or less by accident. This does not in fact correspond to the published data. The only information in its favour was recorded in 1954, when the Sherpas said that in addition to the *mi-te* and the yeti there are other animals that look like Himalayan red bears, but that they never wander into the mountains from Tibet. It is possible that bears do not pass through the mountains, but that does not mean that the mi-te or the yeti does not live in Tibet.

It seems that the whole idea of the yeti not reaching further north than the Himalayas was created by Charles Stonor, but for entirely unscientific reasons, apparently to dissociate himself and the expeditions from "Red China" including Tibet; in effect, to shape the scientific research on the political sympathies of the researcher. If we compare the text of Stonor's memos during the expedition in 1954, many of which are included in Izzard's book, it is clear that Stonor omitted those he had collected with sightings from Tibet. Stonor gives some rather strange arguments for his ideas: "All my informants were convinced, that the small animal could be found across the border in Tibet itself. But we should not forget that the Nepal-Tibet border is a true natural boundary. The slopes facing Tibet are flatter, so they would be a barrier to some animals, although not to others." But one has to remember that the whole foreign interest in Bigfoot began with reports of sightings in Tibet. In 1950–53 Czech ethnographer Nebesky-Wojkowitz collected information on the creature in Tibet and Sikkim. And in Izzard's book one can even read that, when the expedition travelled from England in 1954, scientists, explorers and other experts were sceptical about discovering the Bigfoot on the southern slopes of the Himalayas. And now they say the same about Tibet.

There is no reason to assume that the Bigfoot range does not go further north into Tibet, so how far to the north can we actually find this animal? Is the Nepal-Sikkim Himalayas in reality the southern boundary of a much larger area stretching into Central Asia? The Himalaya area may simply be the area which is the easiest to reach for Western European and American travellers and mountaineers but, because of the physical and geographical conditions, it may be the least favourable place to look for Bigfoot. We will return to this issue further on.

In January 1958 Soviet hydrologist A.G. Pronin stated in newspapers that in August 1957 he had twice seen a creature resembling a "snowman" near the Fedchenko Glacier in the Balyandkiik Valley (Pamirs). This pushed the boundaries of the Bigfoot range further towards the north-west. We are to some extent talking about a great chain of mountains, the great arc of the Himalayas and the Karakorum connected to the Pamirs as well as to the Kunlun Ranges, the Sarikol and the Hindu Kush.

In 1957 appeared two articles in favour of the probability of Bigfoot living in the Pamirs. One article with observations and sketches was published by the geologist A. Shalimov, who had seen footprints in the snow similar to the ones found in the Himalayas. The other was stories from the geobotanist Professor K.V. Stanyukovich about a population of humanoids: stories very similar to the Himalayan mountaineers' stories of the yeti. Almost at the same time an article appeared in the newspapers by the Chinese film director Bai Xin about seeing "Ye-ti" in pairs and the discovery of footprints of these creatures in the Pamirs in China (the Sarikol Range). These articles and the message from Pronin caused a rise of interest in the issue of the yeti in the USSR, and was the reason the Academy of Sciences of the USSR organised an expedition to the Pamirs in the summer of 1958.

Unfortunately, the expedition was not only dedicated to the yeti. It was also a physical and geographical expedition, and a botanical and zoological survey of some areas of the Pamirs. The geobotanist Professor K.V. Stanyukovich was head of the expedition. The summer months were chosen as the season of work. This is the time of year when the zone of eternal snow, which can be found just about everywhere in the Pamir area, is separated from the alpine vegetation by vast expanses of naked lifeless rocks. Despite the fact that several experts in the Pamir area suggested that the main focus of the expedition should be the search for Bigfoot, the upper parts of the Yazgulem and Bartang valleys as well as the Alichur and Darvoz Ranges were chosen for the study. These areas are not part of the Pamirs in a narrow geographic sense, but part of the Badakshan Mountains. Two areas in the Central Pamir region were also chosen: the area around Lake Sarez, and the Pshart and Balyandkiik valleys. These areas are of great interest regarding a completion of the geobotanical map of the Pamirs, and the collection of a highly valuable herbarium. Regarding the zoological aspects of the expedition, there were very few resources. There were no primatologists, and not even experts in general on mammals of the mountains. The use of bait was mainly done in areas where there was no reason to assume any Bigfoot were present. And neither was the selection of sites to be used for stationary observations with telephoto lenses. The search for tracks in the snow proved impractical because it was done away from the mountain passes, and in the summer season when the snow recedes to greater heights and the total surface area is reduced. Members of the expedition did not even see a single snow leopard, and we know they are quite numerous in the Pamir Mountains. Caves and grottoes are also very numerous in these mountains, but they were hardly inspected or studied from a biological viewpoint.

The author of these lines participated in the Pamir expedition in 1958 as Deputy Chairman of the Scientific Council, but cannot agree with the description of the expedition given a year and a half later by Professor Stanyukovich in a collection of essays called "Following the amazing riddle". Fascination cannot replace a proper scientific report on the work of the expedition on the Bigfoot problem. The most cautious would conclude that the expedition did not gather sufficient material for a final judgement about the presence or absence of Bigfoot in the Pamirs, or the possibility of sporadic visits from there to neighbouring Xinjiang (China). Stanyukovich's conclusion, that if archaeologists discovered traces of Stone Age people in the Pamirs it would make it impossible to search for Bigfoot, is wholly unjustified.[42] The idea that Bigfoot could only be found where there have been no people before, and nobody is living there now or has been there recently, is not logical. This whole idea is hopeless anyway, because the Bigfoot is not only found in Tibet, but also in the country of the Sherpas in Nepal. The whole thesis is far too simplistic. The density of the human population in the Stone Age was so small that there is no reason to think that the Pamirs were entirely occupied by people, and that they had displaced the lower primates. Stanyukovich's conclusion clearly shows that he had no biological ideas about what he had promised to look for in the Pamirs. This is clearly confirmed by a science-fiction story published by him after the expedition called "The man who saw it" which describes how a researcher in the Pamirs is stalked by the last living Bigfoot, which in the finale of the story perishes in the turbulent waters of a mountain river. I am not disputing the author's right to write whatever fiction he likes, but the image of the yeti as described here, as a helpless herbivore not able to get away from from its pursuer, is completely wrong. If this is your idea of the yeti then of course nothing can be found in the Pamirs, even if a true Bigfoot was hiding there.

It was therefore wholly unjustified when the Chairman of the Scientific Council of the expedition, S.V. Obruchev, concluded in 1958 that "the environmental conditions of this mountainous country are not favourable for the existence of a large primate". This statement begs the question. Did the leader of the expedition and the chairman of its scientific council not know anything about the ecological conditions of the Pamirs before the departure of the expedition, although it had already been described in specialised publications and in textbooks? The 1958 Pamir expedition did not in fact contribute to the previous knowledge about the environmental conditions of the study areas.

Another case is the treatment of A.G. Pronin's observation from 1957. According to Stanyukovich "climbers and scientists are convinced that the distance from the place Pronin was standing, to the place where he saw Bigfoot, was too great to see anything." In the preface to Izzard's book S.V. Obruchev says equally categorically: "Pronin was at such a great distance from the slope where the animal was moving that it could not be distinguished from a human or a bear." In reality nobody had checked and, as Pronin was not part of the later expedition, there was no way of pinpointing exactly where he was standing and how far he was from the slope.

However, the Pamir expedition in 1958 did to some extent investigate the Bigfoot problem. A small group went to the central Pamirs and interrogated various ethnic groups, following three different routes and collecting more than one hundred eyewitness descriptions.

Three of the researchers (A.L. Grunberg, V.L. Bianchi, B.F. Piston) were of the opinion that the Kyrgyz and Tajik stories of wild hairy men (*gul-biyavan*, a*dam-dzhapaysy*, etc.) are perhaps not a proof of great apes living in the Pamirs currently, but at least that they have done so, but that the memory of these creatures has been turned into folklore. One of the researchers (ethnographer A. Z. Rosenfeld) on the other hand would comment even the most realistic and detailed description by the shepherds and hunters with the words "this cannot be!" and therefore relates to the field of mythology. The last group (A.L. Grunberg, B. Porshnev) went to the extreme south-east of the Eastern Pamirs, where the Kyrgyz guided them into the areas of the "wild people". In general we can see that in the Western Pamirs, the information on Bigfoot is the most obscure, in the Eastern Pamirs they are more detailed, and in the Chinese Eastern Pamirs (south-western Xinjiang) the Bigfoot is described just like any other wild animal.

Thus, the main result of the Pamir expedition was the collection of local material about the Bigfoot. The results were exactly the same as found in the writing of the Mongolian Professor Zhamtsarano.

I had already been trying to synthesise a variety of sources and different directions of research on the unknown hominid primates of Asia. In the summer of 1958, I had already taken a solid step in the direction of putting the whole problem of the Bigfoot on a sound scientific basis. I suggested that the sightings of unknown bipedal primates in Mongolia and other regions of Central Asia (Chapter 8) and the data collected by foreign researchers on the unknown top bipedal primates in the Himalayas and Karakorum (Chapter 4) show they all belong to the same species. This unites the previously isolated sighting areas. It should be emphasised that the fact that the data is comparable is itself to some extent a proof of the reliability of the data, as their origins are quite independent of each other.

It should immediately be stated that this is not just my idea, but that I have also had sponsors, but even so it was wholly justified by the available material. Professor V.A. Khakhlov (Moscow) had also informed the public about the Pamir yeti and also the Tien Shan "wild man" (*ksy-gyik*) as early as 1907–1914. Professor Khakhlov later formulated a hypothesis about likely seasonal migrations of these creatures from the Tien Shan Mountains to the Himalayas, and how they can have been gradually displaced towards the south, i.e. Tibet and the Himalayas. Similarly, Professor Dr. Rinchen (Ulan Bator) was of the opinion that the Mongolian *almas* (or *hsiung-guresu*) is no more than a "close relative of the Mongolian Bigfoot".

To a certain extent, and independently of all these authors, the anthropologist Dr. E. Vlcek (Prague) has dealt with the first reconstruction of the skull of the Himalayan Bigfoot, but has also found data and images relating to the Mongolian hsiung-guresu and the Tibetan *mi-gyo*, showing that, at least in the recent past, they looked similar throughout this gigantic territory.

The French anthropologist Professor Vallois has also studied this issue. He only knows that the Soviet and Mongolian scientists have noted the presence in Mongolia of a being similar to the "classical" one in the Himalayas. But this to some extent only complicates matters, as nothing proves that almas and the yeti are the same. There is also one surprising fact: when we move from the Himalayas to Tibet, the stories suddenly stop, and only come back after another thousand kilometres in Mongolia. This is a strange gap in the distribution. In fact it is only a proof of the author's lack of knowledge, as there are abundant data related to Tibet, Xinjiang, Qinghai, Gansu and other areas of Central Asia (see Chapters 3, 4 and 6).

At the height of the Pamir Expedition in 1958, the newspaper *Pravda* received a bunch of letters in response to the newspaper publishing my article about the Mongolian almas. Several correspondents wrote describing the exact same creature with almost the same name – *almast* and other variations. Later we began to receive information from different areas of the Greater Caucasus as well as from the Talysh Mountains. Based on the analysis of this material and the similarity in names in the various areas, we see there is a mixture of fiction and fact, showing us that the animal was becoming more and more scarce; in areas where it has disappeared altogether, it has lived on in the legends and stories of the population.

Having an idea of the distribution of the Bigfoot there are certain questions we need to answer. What is the nature of this vast territory? What does it tell biologists that could help them reveal the secrets of Bigfoot?

On one hand it is extremely geographically heterogeneous. There are very different physiographic and biogeographic conditions in different parts of the area. On the other hand it is still in some respects uniform.

First of all, it is a geomorphological unit. The central part of the Asian continent, Mongolia, Tibet and surrounding areas is a vast plateau, a place of crossing mountain ranges rising to even greater heights. In the north this region is limited by the Altai-Sayan mountain system. In the south it is bordered by the Himalayas. In the west it is connected to the mountainous country of Afghanistan and Turkestan. Further west there is no continuous alpine zone, but mountains through northern Persia and the Caucasus and the Alpine system of Europe, ending in the Pyrenees. In the north-east the Central Asian area borders the Goltsev region ranges of Eastern Siberia. We can also see earlier biogeographical connection to the north-east and North America, and towards the south in the direction of Malaya and Sumatra.

If we look at the hydrographic map of Asia, we see that we are interested in a territory surrounded on all sides by so-called external drain regions. One river system brings water to the Indian Ocean, the second to the Pacific Ocean, the third to the Arctic Ocean. It was along these waterways that people first settled in these regions. But in the undrained Central Asia, or rather the area of internal drainage, there are no natural ways of human settlement. This negative character also unites the area we are studying. Finally the inland areas of Central Asia are directly adjacent to a second drainless region towards Iran.

If you look at a map of the population of Eurasia, it is also clear that we have delineated an area markedly different from the rest. First of all it is a large and almost uninhabited area, with a population of less than one person per km^2. Only along the edges are small areas with a population of 1–10 people per km^2.

As for the biogeographical features of our area, there is a huge variety of native species, plants and animals. Some of them are only found here, but other species have a much wider distribution, although they are also found in the same area as the alleged Bigfoot. All of this is of great importance to the biogeographical approach to the Bigfoot problem. In our study area one finds for example all five existing species of snowcock: Tibetan, Himalayan, Altai, Caspian and Caucasian. Another animal whose distribution is very close to the distribution of the Bigfoot is the large bearded vulture (*Gypaetus barbatus*). Here we also find several different species of pikas.

The largest predator in the Bigfoot area is the snow leopard. In some areas to the west and other outlying areas, the snow leopard is substituted by mountain living forms of the ordinary leopard. The general distribution of the snow leopard is very close to the habitat of the Bigfoot. These two species seem to be close biological neighbours. Both generally live at a height of 3500–5500 m, often in rocky alpine terrain. The same relationship can be found in the Mongolian People's Republic regarding the almas in Zaalktayskoy and South Gobi. Here there are no high mountains, but the almas meet the snow leopard on the desert plains and groves of oases at a height of only about 100 metres.

The presence of snow leopards in close proximity to the Bigfoot is the solution to an important question. Sceptics often ask why no bones of the Bigfoot have been found, if not on the surface then at least in Quaternary sediments. A major reason is the snow leopard, because naturally there are no bones or corpses lying around. The corpses will be quickly stripped by predators, perhaps even other Bigfoot, and the bones will be removed and crushed by mountain waters. Although one must take into account the weak development of the Quaternary paleontology in mountainous areas.

Another important question is what will happen in the foreseeable future of this area with continued reduction of natural habitat. We propose a study of a more limited area while there is still time: the area where we are most likely to find breeding Bigfoot and the biggest concentration of the animals. This is the western region of the People's Republic of China - part of the territories of the Tibetan and Xinjiang Uygur autonomous regions. In this way ours differs from the map suggested by S.V. Obruchev. His idea, that the main area of Bigfoot is biased towards western Tibet, cannot be considered correct, as in fact the legend of the "snowman" can be found only in the mountain range of the Himalayas and Karakorum in Tibet - "the remaining area is filled with stories about other types of wild man". "Legends like this are known from many other areas of the globe." The problems regarding Obruchev's ideas about "another type of wild man", "other areas of the globe" and so on, are shown in my article "Two openings or one?". The main problem is that Obruchev's concept is solely based on data from Anglo-American studies in the Himalayas and Karakorum. It has already been shown that all this is far too one-sided.

This discussion shifts the emphasis of further research to China. Under the Chinese Academy of Sciences, in the Institute of Vertebrate Paleontology, the authorities have already created an organisation dedicated to the study of Bigfoot in the entire territory of China. The world-renowned anthropologist Pei Wen-chung has been appointed leader of this research program. He has played a huge role in e.g. the discovery of bones of the Gigantopithecus. It is hoped that his extensive experience will help further development in the field of Bigfoot studies. It seems cooperation between Chinese and Soviet scientists is the only way forward.

We must emphasise though, that this project is only devoted to studying the Asian (or, if you prefer, the Eurasian) areas. But surely there are other alleged centres of contemporary or recently living beings like the "snowman" on Earth? Yes, there are theories along those lines. In his book, Ivan Sanderson has already given a summary of the descriptive material of possible relict "sub-humans" from Eurasia, Africa, the Orient (India, Indochina, Indonesia), North America and South America, leaving only Australia and Oceania, and even Antarctica.

PART II

A Review of different Geographic Areas

CHAPTER 6 • DATA FROM TIBET, CHINA AND MONGOLIA

In the previous chapters we have discussed parts of the history. The following chapters are an overview of data from various geographical sectors.

Kashgar

This area can be found on the borders of the Soviet Pamirs, Afghanistan and India (Kashmir), at the junction of the Sarikol, Kunlun, Karakorum and Hindu Kush mountain ranges.

Duvona Dostbaev, 72 years old, born in China, in the Danbash area, has been living in the USSR, in Murghab (Pamirs), for forty years. These events took place around 1912 in China in a place called Tekelik-dzhaylyau, in the Tashkurgan area. He heard the story from his father's brother. One day four men went hunting. One of them, a hunter named Nazar Baatyr, a very strong man, wounded an argali sheep, and followed it up a hill. The others remained at the bottom. He found the wounded argali, killed and skinned it, and hid part of it under a rock, as he could not carry all of it. Suddenly a creature similar to a man, but taller and covered with short hair like a camel coat, appeared. The hunter threw down his gun, battled the creature hand to hand, knocked it to the ground, and tied its hands and feet with a rope. Before the attack the "wild man" had issued a great shout like a man, but he stopped screaming when he was bound. The other three hunters climbed the mountain and found the wild man already bound. One of them was sent to the camp for the yak to carry the creature down. When the yak came, they put the creature on the yak with great difficulty and brought it to their camp. They tried to feed the creature with fried meat, but he would only eat it raw. Next morning they loaded the killed argali sheep and the creature onto the yaks and brought them home. The next morning, they sent a message to Tashkurgan to inform the Chinese authorities about the incident. Several people were sent out from Tashkurgan with horses and carts. They were very grateful to the hunter and gave him lots of money and gifts. Then they took the *gul-biyavan* and brought it to the administration in Tashkurgan.

An old man who lives in a yurt at the Pamir Highway explained that the wild man is both a supernatural creature and an ordinary one, who many have seen in previous years. It is a little shorter than a human and covered with hair. A few years ago, I think in 1946–1948, I met Kudai Vergen. He was working as a guide on the other side of the Chinese border. Early in the morning he was on horseback in a mountain pass, when he collided with a wild man covered with hair. The horse was scared as was the rider who, without waiting for the approaching wild man, took to his heels, but, when looking back, allegedly saw that the wild man ran after him.

Mamedniyaz Aytmambetov, seventy years old, a former hunter, said that he had heard of the wild man (*adam-dzhapaysy*) since childhood. Just as there are wild yaks and wild camels, so there are wild men. They can be found in China in the Rask [43] area. They are like people, but have no clothes and are covered with hair; they are afraid of humans and run away when they see them.

Ernazar Dzhonboldyev from Koytezeka, 74 years old, said: "I have heard many stories about the fact that the wild man can be found in China. There are also a lot of other living creatures there: bears, foxes and so on".

Aydarkul Kudbaev, shepherd from Këldzhilga, 64 years old, had heard that in some areas of China there is a wild man which eats grass and walks around naked. Hunters will hunt him with dogs.

The driver of the Pamir Biological Station in Chechekty, Sultan Tashtanbekov, 38 years old, said that he has heard the "wild men" can be found in Raskeme and Kulan-Aryk in China. They are like humans but the body is covered with short hair. Locals hunt the wild man and eat his flesh. It should be very tasty. If they capture him alive and hold him in captivity he will not eat, and he does not speak.

Mamedkerima Baktybaeva, 68 years old, who was born in China, but moved to the Soviet Pamirs in 1944: "My grandfather lived in Kulan-Aryk in the Rasko area, from him I heard that there are wild yaks, wild horses and wild people. The wild man is like an ordinary person of the same height, but covered with short grey hair. He feeds on grass and plant roots. He lives in hollows that he digs in the ground. He is very afraid of humans, so if a person comes close he will take his young and run, but it can happen that he will attack a human. He is found in the area near Sanji Kulan-Aryk - this is a great valley, where there is water and grass, but no forest, the place is deserted, and not inhabited by people. Whether he lives elsewhere I have never heard, but right now it is a local law that it is forbidden to hunt or shoot the wild man. When I lived in Kulan-Aryk this happened. One day, while working in a field, a man saw tracks of *gesu tyrmak* (the terms *gul-biyavan* and *gesu tyrmak* are used by Baktybaeva as identical, in the sense of 'wild man'), and spent the night in a yurt at the place where the tracks were found. At night the gesu tyrmak approached the tent, but the hunter shot him. In the morning he and three or four people tried to track the wild man, but lost the trail of blood." Baktybaev has no doubt as to the existence of the wild man, but speaks of him in the past tense. "This is not because of drink, he really existed once."

Renowned Murghab hunter Khalbaev Kurbanov, 44 years old, was born in China and lived there for 16 years. He says that over the border in China is a wild man (*adam-dzhapaysy*). Two people who came from China to Murghab confirmed this, both were old hunters. They said that the wild man can be found in the Itarka area, that these creatures live on the plain where there are reeds, thickets, trees and sand. They look like a man, but are covered with short hair. When they see a hunter they run away from him on two legs.

Shepherd Elchiev Qalandar, from Aydynkul gorge, 54 years old, said that according to the local population in the Parang Raskeme area in China there is a wild man living in the area. His height is like a normal man, but he is covered with short hair, and lives in wooded areas. He avoids all contact with humans. He feeds in wooded areas. Man, he never touches: on the contrary, he avoids him. They say he feeds on wild fruit.

K.V. Stanyukovich: "The first information about the existence of the *yavana* I received in 1935 in Kyzyl-Rabat (extreme south-east of the Pamirs) was from one old Kyrgyz, who told me that when he was a young man he was in Togdumbash-Pamir (Xinjiang). Here he had to leave a valley with good pastures, as a wild man had turned up. It dragged his sheep away, and scared everybody."

The Chinese film director Bai Xin from the studio of the People's Liberation Army of China. He said that in 1954 he and three of his colleagues were on Mount Muztagh Ata at a height of 6000 metres. One morning after sunrise, at a distance of about a thousand metres, they saw two "men" - short, with hunched backs, following each other up the mountainside. The creatures, ignoring the men's screams and gunshots, continued climbing upwards with ease, and then disappeared among the rocks. Later Bai Xin and a photographer discovered large tracks of a bipedal creature in the snow. Bai Xin and his colleagues followed the trail for half a kilometre and found a few drops of blood on the road. The tracks finally led them to a massive block of ice, where they had to turn back because of the lateness of the hour. Bai later learned that among the population and border guards there are a lot of stories about "wild people" in the Chinese Pamirs. At one guard hut the guards had thrown some cows meat about 40 metres away. In the moonlight they saw a white-haired wild man grabbing the meat and running away with it. Based on this information and their own observations, Bai Xin is convinced that this creature really lives in the mountains of the Chinese Pamirs.

Major-General P.F. Ratov reported to the Commission that, working in Xinjiang in 1939–1940, he received reliable information on "wild people" in the three districts of the Chinese province, especially near Tashkurgan. "Chinese soldiers, who served in Tashkurgan, met the wild people many times. These creatures do not speak. If you try to move closer, they run off. The same happens if you pull food on a string. But if you leave it on the ground, sometimes they will come closer and take it. Everything suggests that, in this area, there may be a large number of wild people."

In May 1959, the geographer B.A. Fedorovich recorded the story of Mattuka Obderaima, 26 years old, who worked as a policeman, but who is now in a commune in Tashkurgan. In 1944, when the narrator was 11 years old, he went to visit his uncle Nuruz Muhammad, who lived in the Rusk-Darya, a three-day journey on horseback south of Tashkurgan. During his stay his uncle brought a freshly killed wild man (*Jawo-halga*), back from a hunt. According to Mattuka, it was a creature more like a man than a monkey. The skin was covered with thin and short yellowish hairs. The colour and thickness of hair on the back and front of the body are the same. Jawo-halga hair is different from a bear in that it is not so fluffy and more short. The hair on the head was 15–17 cm in length and the same colour as the coat, the face has no hair. There is no tail. The feet of the jawo-halga are wider and shorter than in humans. The tracks are also very different from the much broader bear tracks. In the hands the thumb is closer to the other fingers than in a person. Mattuka did not know whether his uncle's family kept the skin of the animal. He had not been there for six years. Many in the area knew of the jawo-halga. When this animal meets up with humans, it quickly runs off into the mountains on two feet, and with the speed of a mountain sheep. If it is persecuted, it often turns around and makes a sharp guttural cry like "ee". It is unlike all other creatures. According to Mattuka, the jawo-halga is found in areas where the natural conditions are less severe than in Tashkurgan. There is much more vegetation as well as wild camels and wild yak.

Professor Fedorovich's survey also interviewed Hadyrbek of the worker's committee of the commune near Tashkurgan. According to them, the wild man (*java-adam*) lives on the border of Xinjiang and Kashmir where there is a lot of water and good vegetation. Stories of them are also told in the mountains and adjacent areas of Afghanistan and Pakistan; it is found, allegedly, in the mountains above the Karglyk village. Hadyrbek told about a meeting with a wild man in 1941–42. In the frontier area hunters of wild horses and wild yak saw the wild man on several occasions. They do not hunt him.

Finally, we have information received through Prof. Pei Wen-chung of the Xinjiang branch of the Chinese Academy of Sciences. They are also concentrated in the area south of Tashkurgan.

In 1954 a "man-bear" was caught in the Upulan og Taban mountains (south-east Tashkurgan). Its face resembled a monkey, its arms were bent like a man, and it had opposable thumbs like a human. It had very sharp nails, and the hands were big enough to cover and throw relatively large stones. This "man-bear" is totally different from an ordinary bear. The legs and the tracks are very human, and so is the body position. It walks on two legs, but on all fours when pursued. In the Shikan area (south-west of Tashkurgan) there is also a remote and wooded area with ice-covered mountains inhabited by groups of "wild people". They are covered with brown fur, which is reported to be shed in April, they are very agile and walk upright, they hold their children in their arms like a human. In his letter, Professor Pei Wen-chung writes that they are researching these reports.

There is also a report from Professor E.M. Murzaev about some local hunters claiming that, on the border between China and the Soviet Union and India, they had met a *yabalyk-adam* that threw stones at them.

Tibet

A whole group of information refers to the area around the Rongbuk monastery on the northern slopes of Mount Everest, as well as all of the territory which lies directly to the north of the Nepal-Tibet border in the Himalayas. As mentioned above, in 1922 a senior lama of the Rongbuk monastery said that five *e-gyo* lived in the hills above Rongbuk. Norman Hardie said that to his knowledge this monastery houses a yeti scalp, like the ones found in the monasteries in Pangboche and Khumjung.

Here are the stories of two of the Tibetans Stonor interviewed in his book. They came from a village near the Rongbuk Monastery. Two or three seasons ago in the early spring a few boys from the village, who were grazing their herds of yaks, saw a short man walking towards them, but as he drew near they saw he was not a man, but a yeti with a pointy head and a body covered with dark red hair. The boys took to their heels. Another well known incident in the same district happened a few years ago. After an unusually heavy rain, a mountain lake overflowed and poured water down the slope. When the water disappeared the body of a yeti washed up on the rocks. Many of the villagers saw the corpse, describing the yeti as a short man with a pointed head and reddish-brown hair, but were afraid to save his skin. The monks from the Rongbuk monastery allegedly often see the yeti in the winter months, when they come down to lower heights. The cry of the yeti is familiar to most mountain people and can sometimes be heard near the villages. The Tibetans believe that this animal eats small animals, probably pikas living among the rocks, and in the summer large insects.

All Nepalese informants were convinced that the yeti could also be found across the border in Tibet, and many of the informants were Tibetans who came from the other side of the border. One of them was the merchant Nima Pasang, who arrived with a caravan from Dvingr village a few days journey north of the Nepalese border. He said that he saw the yeti in September or October of last year, when he together with other pilgrims climbed into the mountains to a sacred place for religious ceremonies. When they arrived, other Tibetans told them that they had seen a yeti not far away. Pasang and his friends went to see the beast. And they did see it, among the rhododendron bushes at a distance of some 200–300 metres. The creature was the same height and build as a small man. His head was covered in long hair, as was the middle part of the torso and the thighs; the face and chest of the beast seemed less hairy, and the hair on the legs below the knees was short. The hair on the chest was reddish. The creature walked on two legs and would stand erect almost like a man, but it was also bending down all the time, apparently in order to dig up some roots. They watched him for a long time, but he never went completely down on all fours. After a while the yeti noticed the people, let out a loud shriek, rushed off, still moving on two legs, and disappeared into the undergrowth. After his escape, Pasang said, we walked to the place where he had been, and saw the footprints in the soft soil.

Another one of Stonor's informants was a Tibetan who arrived from across the border. He described the animal as walking on two legs, being stocky and having the torso and legs covered with long hair, although the breast was less hairy. The colour was like a *Tahrah-gornogo* goat. The informant, Pinechu, claimed he had heard the scream of the creature many times. He even reproduced the cry. The creature eats small animals it finds among the rocks. It waits for the animals to come out of their holes, and then grabs them and kills them by bashing them on the rocks. Then it will pull out the insides of the animal and discard them before eating it. Pinechu said that he had once spent a night in the mountains, and when he suddenly woke up, he saw a beast in the moonlight squatting on its haunches a few metres from the fire, that the people had extinguished. Overcoming his fear, Pinechu threw a piece of resinous wood into the embers of the fire, which made the beast run away. Another resident of the same village had told Pinechu that he had seen a beast walking on its hind legs like a man along a path. The villager hid behind a rock and saw the animal go off the path and start to climb the boulders. He found its prints in the fresh snow. They looked like the prints of bare human feet.

The late eminent Soviet Tibetologist Prof. Y.N. Roerich, who lived for many years in the Himalayas and Tibet, gathered abundant information about the creature known as Bigfoot. We will return to Professor Roerich later, but for now note one single passage: "Not far from the Nathu La pass, on the Tibetan side, there is a Buddhist monastery, Dungkar, at quite a high altitude - about 3000 m. Local monks say they constantly have to deal with this creature, which in their eyes is not conjecture or some dark rumours, or blind tradition, but a real fact. When they go into the forest to fetch wood and collect grass, they have seen 'wild men' many times."

We have already reported on the Chinese Academy of Sciences research on the problem of Bigfoot, where they have used questionnaires to collect data in the Everest region (Dzhumulama). Local residents who claim to have seen the *E-bian* (literally "man-bear") are very well versed in nature, and know the different species of bears and their tracks. In addition, the Chinese researchers managed to get a sample of hairs allegedly belonging to *E-bian*; the study of this sample is not finished yet, but the hairs are very thin and soft, and look nothing like a bear. They seem to belong to a member of the primate order.

Zoologist Xu Wen gave a report on these and other results at the Conference on Paleoanthropology at the Chinese Academy of Sciences in December 1959. In the Chinese press Professor Pei Wen-chung reported that "the scientists have reported on new material about Bigfoot in the highlands of Tibet, and this has expanded our understanding of the enigma of Bigfoot."

There is a mass of information about Bigfoot in countries to the south of Tibet, but we also have information from countries that lie far north of Tibet. If experienced hunters cannot find the wild man in areas inhabited by Kazakhs, they will move deeper into

Central Asia, heading for Tibet. The lama of Agin Datsan (Buryatia) warned Valery Grachev that he could only follow a caravan of pilgrims to Tibet if he was not afraid to meet the wild man. An interesting letter from J. Zhamtsaranova says it is not uncommon for pilgrims to meet the "man-bear" on the pathway to Agin Datsan. Her mother had told her about the creature. She had described it as we have heard it before, but adds an interesting thing about the creature's habits. It is covered in hairs and walks like a man, but carries a stick, which it uses to dig for marmots. Pilgrims often saw it digging. They would hide from it, but it would not attack.

People far to the west of Tibet have heard the same thing: the Pamir Kyrgyz hunter Turganbaev Shaimkulov said that according to old people in Tibet there is a wild man (*adam-dzhapaysy*); it looks like a man, but when it sees a human it runs away.

Tokoev Kadir, who works on a yak breeding state farm in Bulun-kul in Pamir, said that when he told his father about a meeting with the gul-biyavan in 1922 or 1923, his father did not believe it. "He said that the gul-biyavan cannot be found in our area, but only on the Chinese side, where there are a lot of them. My father was a very knowledgeable man, a hunter. And my grandfather was also a good hunter. They also told us that in Tibet people in caravans often see wild men. He told us there is a mountain in Tibet where a whole family of these creatures can be found, and where the local people are afraid to walk, because the gul-biyavan throws large rocks at them and screams."

One of the farms in the Kirghiz SSR is home to about 30 immigrants from China, who, during the war, lived in remote areas of Tibet, earning their living by hunting argali sheep and other game. One of them, named Omraly, said that one of their women was captured by a *Kish Kiik* ("wild man"). It is not much different from a human, it is just covered with hair and cannot speak. When the woman escaped, they told the authorities where to find the lair of the wild man. Several people were sent to capture the creature. According to the informant several wild men and one wild woman had actually been caught, but he did not know more about that.

Several people have collected information about the wild man in Tibet. Some stories were recorded by the Czech ethnographer Nebesky-Wojkowitz in 1950–1953. One of the informants was a Tibetan official named Niameh. He had received a Western education and was not superstitious. He claimed to have seen the *mi-kyo* three times. The first time when he was a child in the Dungkar monastery, the main monastery in the Chumbi Valley. One night the inhabitants of the monastery and Niameh had heard the characteristic whistle of the mi-kyo. The next morning they had found tracks in the snow near the monastery walls. The tracks had disappeared into the thicket. During a snowy night many years later, Niameh and his companions had been in the Nathu La Pass in the Himalayas. Here they had heard the whistling cry of a mi-kyo near their camp. At dawn they found its tracks and went out to find it. A mi-kyo suddenly appeared from behind a rock. It was an apelike creature with a rolling gait.

It departed without haste and disappeared behind a mountain ledge. Niameh met the mi-kyo for the third time in the forest in Chumbi Valley. It was late at night, and Niameh and his companions heard the sound of an animal approaching the camp. They quickly fanned the fire, and saw a mi-kyo circling the camp for some time.

Nebesky-Wojkowitz has recorded another interesting story about a Bigfoot captured near the Nathu La Pass. This story is all the more interesting because Prof. Y.N. Roerich has heard the same story from another source. The men in a Tibetan guard hut near the pass had noticed that a mi-kyo would come to the house and drink water from a wooden trough. This happened several times, so they decided to pour Tibetan beer into the trough instead of water, and capture the creature. So they filled up the trough, and then hid themselves, armed with ropes and poles. When the mi-kyo came back, it started to drink the beer. It clearly enjoyed it, because it drank all of it very quickly, and then fell down drunk. The Tibetans rushed up, and tied its hands and feet to a pole. And then they started to walk off with their prey, and try to find a European that would give them a rich reward for their rare catch. They did not get far. The mi-kyo suddenly woke from its intoxication, broke the ropes and disappeared among the rocks in long jumps.

The story recorded by Roerich diverges in some detail, but is basically the same. "When I was in the Nathu La Pass area, I met with a group of Tibetans who were with a caravan from Tibet. They had crossed the Dzele Pass and later arrived in Kalimpong. They had heard of the Europeans' interest in the yeti. The group told me the following curious story. After crossing the pass, they spent the night in the wood at an altitude of over 4000 m. As always they had *chang* (Tibetan beer made with barley) with them. When they went to sleep, they had left a container with some *chang* near the fire. One of them woke up at night and saw a creature which they call *mi-gyo* approach the fire. He kept still, thinking that this was a golden opportunity to see what the mi-gyo would do, and maybe catch it and profit from it. Everyone says that the creature is very curious when it approaches a camp. It will move objects and seems very interested in things it does not normally see. When the mi-gyo found the container with the remnants of beer, it tasted it, and apparently liked it, as it drank more of it. Then it got drunk and fell asleep near the people. The sleeping mi-gyo was captured and tied up. Its arms and legs were tied to a pole, and it was dragged down from the mountain. They hoped to carry it to the town of Kalimpong. But on the way the creature came to, and because it is very strong, it broke the ropes and escaped. It happened around 1938–1939."

There is also a third version of the same story given by the Greek anthropologist Prince Peter. It differs from the other two in a few minor ways, but is basically the same.

In the book *Tibet is My Home*, Tibetan author A.M.F. Tsevang has written the following on the yeti: "Since early childhood I have heard stories about the mi-gyo.

The area around Yatong (Chumbi Valley) is considered to be its favourite place, especially Chumpithanga on the road to Sikkim. It is believed that the mi-gyo is a huge creature like a monkey. It will eat fruit and berries, and it lives in the mountains. In cold winters the mi-gyo will move down the mountains. It is said that the mi-gyo is always attracted to the smell of roast meat, so newcomers to Chumpithanga are urged not to fry their meat. Mi-gyo also has long hair that cascades down from the head. It is a sacred animal, but it is considered unfortunate to see it." Tsevang also repeats some stories about people having tamed the Bigfoot. Monks from Geshe Rinpoche Dungkar are supposed to have tamed a mi-gyo and used it for carrying fuel, water and food. Tsevang is apparently not sure about this story, as he continues: "I know only one person who has seen these ape-like creatures himself. He was an honest worker, without much imagination or flights of fantasy. Once on the road to Yatong, he passed near Lake Tsomga 20 km from Chumpithanga. The day was cold, it was snowing. Suddenly he heard the sound of crunching branches coming from the woods to his left. He looked towards the sounds, and saw a humanoid about 1.5 metres tall with a red face and red hair, standing at the edge of the forest looking at him. Then the creature ran across and disappeared in the thicket. When the man arrived in Chumpithanga he learned that many of the locals had seen the mi-gyo."

Another Tibetan, Ge Lan, says that Tibetans are in no doubt of the existence of the mi -gyo. Everyone knows about it. He had personally seen animal parts, particularly skull caps, which are attributed to the mi-gyo and used in some ceremonies. It was Ge Lan's impression that the skull caps he had seen in the hands of the monks were much longer than a human skull cap. Ge Lan had also heard that other monasteries have skins of the mi-gyo. According to the stories of hunters, the mi-gyo is a carnivorous predatory animal, although it also eats various roots. Ge Lan has seen hunters bringing scraps of hare carcasses down from the mountains. These are credited to the mi-gyo. According to the stories, the mi-gyo is covered with hair – it is brown with a sheen, thick, but not long, except on the head, where it is very long. Ge Lan knew the hunter, who said that in the upper part of the body of mi-gyo the hair is growing upwards, but on the lower part it is growing downwards. The body of the mi-gyo is like a monkey, but a little taller than a man. The face of the mi-gyo is also both like a monkey and a man. It gives a loud cry like "poo-ku, ku-pu". According to Ge Lan, the mi-gyo not only lives in the south of Tibet, but also in the north part in an area he called Nari (Ngari).

The Polish journalist Marian Belitskaya, who visited Tibet in 1956, has also collected interesting records of the mi-gyo. During his stay in Lhasa, he learned from high-ranking lamas that the term de-ti (in Nepalese) and mi-gyo (in Tibetan) means the same. Later he learned from his interpreter that the Tibetans are convinced that the mi -gyo exists, that it lives somewhere in the Himalayas and the more northerly mountains. Belitskaya also found a man who was not afraid to talk to a stranger about the mi-gyo and his close encounter with one. "Two years ago I was passing through Yatong with my brother on the way from a nearby village, when I heard crackling in the bushes and saw a mi-gyo peeking from the branches. The mi-gyo was brown, with a lot of hair on his forehead. He did not produce any sounds, He also had long arms covered with hairs." The local people claim that the mi-gyo come down from the mountains into the valley, but that they normally live only on the highest peaks. They live in caves or in inaccessible canyons. It is rare for them to come down. They only come for roots and fruits, and sometimes in very cold weather.

A lama of the Tashi Lhunpo monastery in Shigatse told Belitskaya that monks who lived for long periods in seclusion in the desert often met with the mi-gyo. Some were able to tame them, and used them to bring branches, food, water, edible roots and berries. The creature looks like a tall human with the body overgrown with light brown hair. It lives in the high mountains, where it eats roots and berries. It moves on two legs, and its feet are larger than a normal person. According to these monks, the mi-gyo is not aggressive towards humans. Local residents however believe that mi-gyo sometimes abduct people, and than you should run rapidly down the mountain if you meet one. One hunter claimed he had once shot a mi-gyo. An angry mi-gyo only emits inarticulate sounds, and the animals only communicate with inarticulate sound. Belitskaya could also report an important result. Only very few people would speak freely with him. Several lamas confirmed that they had heard about the mi-gyo, but also that it was not proper to talk about it with a stranger. It is considered bad luck to tell a stranger about the creature. One prominent lama told Belitskaya there were two reasons for this. One, that the people do not like to talk about sacred things, and that the mi-gyo is considered sacred. Two, that if news of the mi-gyo reached distant lands, it would bring foreign expeditions, and they did not want that.

V.O. Wang, who is a Chinese student in Dnepropetrovsk, has written about the mi-gyo. In many places in Tibet people have summer and winter homes, and will move with the seasons. Some of these people have seen the mi-gyo and one allegedly found two dead bodies next to an empty container of beer. People carried their heads to one of the temples, where the skulls are kept to this day, but the lama does not allow strangers to visit.

This information clearly shows that the animal can be found in the south and in the north of Tibet. There is also a lot of information about the animal in the east of Tibet. This next information comes from the very reputable and distinguished lama Tsultun Zambia who was told, during a pilgrimage to the mountains on the eastern border of Tibet, that a pious hermit had tamed a mi-gyo. According to his description it was an apelike creature almost 2.5 metres tall, covered with 4 cm long hairs and with a dark complexion. The lama said: "He not only did not hurt me, he helped me in my hermitage. He could not talk, but had a great mind."

We also requested information from Sherap Dzhamtso, the chairman of the China Buddhist Association. In his kind reply he told us among other things that the Tibetan mi-gyo is an animal like a monkey, but larger. There are various kinds of mi-gyo and they differ in their growth and strength. Sherap Dzhamtso emphasises that he does not know what kind of animal the Nepalis call yeti, but believes that it is different from the Tibetan mi-gyo. According to Sherap Dzhamtso, they are "like great apes, and they do not understand human speech. They can be found in wooded valleys in eastern and southern areas, but I do not know whether they can be found in other areas…"

One of the most important conclusions based on the Tibet material is, that the lama clergy is a substantial obstacle for research in the mystery of the Bigfoot, although they seem to have extensive knowledge of it. There are reports that in some Buddhist temples and monasteries bones, skins and even entire mummified mi-gyo are used for worship. According to Lama Chemed Rigzin, he has examined the mummified bodies of two mi-gyo: one in the Riwoche monastery in eastern Tibet, the other in the Sakya monastery in southern Tibet.

Stonor also reports that an old man told him there is a monastery near Shigatse in Tibet, where he had seen skins of bears, leopards and other wild animals, as well as the hide of a mi-gyo.

Finally Valery Grachev has reported he is aware of the presence of a mummified mi-gyo among other animal mummies stored in Sakya in southern Tibet near the Trom-chu River.

Yunnan

We have a limited amount of data from the mountainous western part of Yunnan Province. In one of his letters, Professor Pei Wen-chung wrote: "I know that in Yunnan people say, there are wild people living in the mountains." According to a former commissioner from the Society for Cultural Relations with Foreign Countries, he was told by senior officials and scientists in Yunnan that at the beginning of 1954 some peculiar people were discovered in the mountainous region of western Yunnan. They did not belong to any of the nations. In general, they seemed to be in a prehistoric stage. They lived like animals, hidden from other people. They do not speak, and it was mentioned that their bodies were covered with hair. One of these "people" had been captured and taken to Kun-min. It tried to put on human clothes, and seemed to like it, as it was smiling. According to some reports, the wild man was sent to Peking for further study.

Prof. Mao Fu-chi, deputy director of the Institute for National Minorities at the Chinese Academy of Sciences, has reported on the popular beliefs in the provinces of Fujian and Yunnan. Farmers going to work in the mountains would, according to ancient custom, wear a sort of tube of bamboo on their right hand. If they were attacked by a wild man, they would reach out the hand protected by the bamboo tube. The wild man would seize the tube with his teeth, and the peasant would free his hand and flee. Prof. Fu-Chi adds that similar legends are prevalent in Tibet and Sikkim. Similar beliefs are mentioned twice in the ancient book *Shinhan-jin*.

Finally, there is an absolutely identical tradition in the border areas between Vietnam and Cambodia.

Xinjiang, Gansu
As I have mentioned before, Major-General P.F. Ratov reported to the Commission on the three areas of this region where, according to his information, "wild people" were living.

"Tupan Urumqi told me not to go to Lop Nor, as there are a lot of 'wild people'. They even say that they can eat you. The area is a colossal lowland completely covered with reeds, 450 km in length and 150 km in width with no population." According to Ratov, in 1937 the late Soviet Marshal P.S. Rybalko [44] was driving east along Lop Nor and the north ridge of Altyn-Tag. During the trip a Chinese officer told him that a wild man was caught by a cavalry baggage train. Rybalko found the prisoner lying strapped to a cart. He was told that the soldiers sometimes allowed him to go out on a leash. He always walked around on two feet, and never got down on all fours. Rybalko talked in great detail about this creature. It had no clothes, had a very dirty yellow colour. It looked like a wild human, possibly a fossil ape-man. The hair on the head was very long, going down below the shoulders. He was stooped, with long arms, and only made sounds like squeaks and meows. According to the local population, these creatures live on fish. Rybalko asked to have the wild man transported to Moscow, but it died near the town of Kurlya.

Prof. Burhan, director of the Institute of Languages of National Minorities of the Chinese Academy of Sciences, has informed the Commission about this sighting he received from Saifuddin Azizov, the chairman of the government of the Xinjiang Uygur Autonomous Region. "In 1957, a Uighur hunter met a creature called the 'man-bear' near Lop Nor. The creature moved on two legs, and was covered with dark brown hair. The hunter shot the 'man-bear' and then took off the dead animal's skin and gave it to the city of Kurlya where it is stored at the moment with the Tursun Israil, the district chief." Further research into this case has established that the skin is in fact of a brown bear.

The local people living in the Lop Nor district know about the "wild people", but they are not aware of the fact that the animals are of interest to science. They do not hunt them, as their fur has no undercoat, and so is of no value to them.

It has been told how N.M. Przewalski received a large amount of information about the *hsiung-guresu* ("man-beast") from the people in Gansu, especially in the mountains of Nan Shan, and how later his Cossack Yegorov saw some of these creatures on the southern slopes of the Nan Shan. Many years later, in 1957, the Soviet journalist, retired colonel S.G. Kurzenkov, was in the territory of the Tibetan National District in Gansu Province near Tendzu. The local residents told him that in the remoter areas you could find human-like creatures, which the Tibetans called the mi-gyo. They are covered with thick hair, do not have clothes and walk on two legs like humans.

Another valuable message comes from the Tibetan National District of Gansu province and Professor Jim Peng from the Institute of Languages of National Minorities of China: "Our Tibetan student Ahn Chi-hsing has said that eleven years ago, in 1947, when he was in the fourth grade in primary school in the village of Chzhonisyan in the Luchzhugun area, a mi-gyo was caught close to the Shauvasy temple . Many people went to the temple to see the mi-gyo. They said that the mi-gyo looked like a man, but that his body was covered with brown hair, and that it was very long on his head. A few days later he died. The mi-gyo's skin was removed and deposited in the Hufashendyan temple, where it is supposedly still to be found. The Shendinsy temple in Chzonisyan used to have two drums made of mi-gyo leather, but in 1929 there was a fire in the temple, and they were burned."

We have also received the following from the prominent Chinese zoologist Prof. T.X. Show (Institute of Zoology, Chinese Academy of Sciences): "A few years ago I received a letter from an elderly person, a former army officer. Once, when he was with the troops on the border of Gansu and Qinghai, he saw a wild man in the woods. It was moving very fast, and disappeared very quickly."

Major-General Ratov reports that there is a third area with a population of "wild men" in the Xinjiang region around the eastern spurs of the Tien Shan Mountains in the vicinity of Bogdo-Ula. This mountain is surrounded by lakes and enormous forests, and there is no permanent human population there. The area around the lakes is said to be inhabited by wild men. Officials from the Chinese military confirmed these reports when interviewed by Ratov.

In the newspaper *Kazakhstanskaya Pravda* we have also found information from P.I. Chumachenko. According to him, in 1914 a wild man was caught in a thicket near the Manas River. The creature was delivered to the house of the rural district chief, and from him he learned the following: The height was approximately like a 14–15 year old teenager, feet and hands were similar to humans, it walked on two legs, and only very rarely on four. It was a male, and it was all overgrown with hair – it was fine and thick, short and with a blue-grey colour like the coat of a two-three week old camel. The head and the face were similar to a human, but the forehead low. In the night he was tied with a rope around the neck in the tent. The next day the Kazakhs heard the news of the war with Germany. They decided it was nothing to do with the wild man, so they took him back to the river.

Mongolian People's Republic

Moving east, we now come to the Mongolian People's Republic. As already mentioned, the main body of observations here belong to the school of Prof. Zhamtsarano, and later Prof. Rinchen. They have shared their information with the

Commission, and it will be briefly summarised below. But it should be added, that very much of the recent collection of information has been done by the zoology professor G.P. Dementiev.

As already mentioned, the collected data show three areas of recent meetings with the almas:

1) In the west of Mongolia, in the Khovd region;
2) In the south of Mongolia, in the least accessible areas of the Mongolian Altai and the Gobi desert;
3) In the extreme east of the MPR, in the area around the Khalkhyn Gol River. Of these areas, the Gobi is of extreme interest, since this is where we have evidence of a series of meetings with almas cubs. So far we know of only one other area with evidence of this kind, which means it is probably also a breeding area – the south-west corner of Xinjiang.

According to Dr. Rinchen you can still find old people of the Gobi who have seen the footprints of the almas, but all agree that today you see the almas less and less. Almas have a somewhat stooped figure, the legs are slightly bent at the knees, and the hands are longer than in humans. The hair is long, and the body is covered with sparse dark hair. The females have long breasts. Almas do not know the use of fire, guns and weapons. "It is impossible to deny the existence of our distant cousins so far behind us in their development, and that they are living their last days like wild horses and wild camels in the vast Gobi."

In the northern part of the MPR the name almas is almost unknown, or is used for an unreal, mythical creature or evil spirit, perhaps something like a witch. In the east, south and far west of the Republic, the name almas is commonly used, sometimes *hong-geresu* or *zagyin emgen*, but this is especially for the females.

Here is an example of a recording made by the late Prof. Zhamtsarano: "One day in the village of Copper, at the end of the Mongolian Altai, all the Mongols were hiding in their yurts during a snowstorm. Suddenly, all the dogs started barking like mad, and a group of the nomads came out and scrambled up the hill. They said they had seen a naked hairy man in the raging storm. They tried to call him: 'Chikembi? Who are you?', but when the mysterious man saw the people he ran away. Next day, they found his track in the snow. It was human-like, but also different, he had clawed bent toes, and the first toe had an unnatural size. The tracks went through the steppe to the mountains and disappeared completely in the inaccessible gorges."

Dr. Rinchen also quoted another entry which was personally familiar to him. It came from the old lama of Agin Datsan on the way to Tibet. An *almaskoy* was apparently caught in an area south of the border of Mongolia, but it managed to escape.

When it comes to observation of young, we have the following. An elementary school teacher in the Gobi-Altai in 1930 said that he had seen the corpse of a hairy naked girl about seven years old.

The monk Dambayorin has said that in the 1930s he went to Sugino Toby at night on a camel and saw a naked child. Thinking it was a lost child the monk came closer, but found the child covered with red hair. He fled in terror from this young almas. In the same year another man met a naked baby almas in the same area. He fled too, thinking that meeting a young almas was a bad omen. Some herders have also seen young almas approaching their herds of sheep and have fled, leaving their herd behind.

There is an abundance of information on young ones, and it corresponds with a considerable amount of information about females, in some areas which are probably breeding grounds. The males seem to be more migratory. In Mongolia the female is almost more common than the male.

According to Rinchen, a Mongolian named Anuhu and a companion were driving camels in the southern Gobi when they noticed a thickset shaggy biped running away from them. Both decided to try and catch the runaway on camels. But when the fast Gobi camels caught up with the creature, they saw it was covered with short hair. The beast then issued a sharp cry, that frightened and stopped the camels. And then the *almaska*, as they are called in that area, ran back to the stone ridge, scrambled up it with the help of arms and legs, and disappeared.

Another story tells of a mother coming back to the yurt and finding a naked woman covered with sparse reddish hair breastfeeding the woman's baby. The hairy creature jumped up at the sight of the woman and ran outside, where it disappeared in the saxaul thicket. It had curved legs and long arms.

Prof. Rinchen and Prof. Dementiev have confirmed the presence of strange structures in the southern Gobi region. Whether they are natural, or man-made, or made by the almas is a matter of some discussion. Locally they are called *almasyn dobo* or “almas hills”. Pictures sent to the Commission of one of them shows that this hill has almost vertical walls, like a Mongolian yurt, the top covered with saxaul and the whole structure has variable height, and is spacious enough for a creature the size of a human. The total size of the hill is 3.5 m tall. “The head of the local council, Tsoodol, has said that he has found rabbit bones and feces like human inside these artificial caves or burrows. Now they have been uninhabited for several decades, but according to the old people almas used to live in them. Many of them have collapsed or have been buried in landslides. Places with similar names and burrows can be found in the south-western Gobi in several areas.”

The German Schiltberger mentioned stories from the fifteenth century in old

Mongolian chronicles about how almas in the Gobi lived in earthen burrows.

Additional surveys of the "almas hills" have confirmed that the inhabitants have long abandoned them. Many are partially collapsed. Old people living in the area remember them going hunting at night in small groups. The head of the local council said he removed an old skull from one of the caves, and it was different from human skulls.

In his article "A note on the Bigfoot in Mongolia", Prof. Dementiev reports additional information about the expedition, undertaken in June 1959 by the local headman named Tsoodol, to study specific areas in the Gobi Desert called *almasyn dobo* or *almasin hohhot* ("city of almas"). This sandy area is currently not populated, but Tsoodol collected some information from elderly people who lived here 60 years ago, and personally inspected some of the caves almas had dug in the sand dunes. Unfortunately no excavations were carried out. The Mongol old-timers claimed that 3 –4 km away there used to be a large society of almas. According to the information collected by Tsoodol, at least four almas lived here 60 years ago. They were nocturnal, bipedal and completely covered with thin wool. Their hair was long and thick, and they had an almost hairless face, where the skin had a bluish tinge. Dementiev added that neither bears nor wolves are found in this area. But it is impossible to know whether they lived here 60 years ago.

There is some information about material remains of almas in Mongolia. However, none of it has been examined by scientists. According to Dr. Rinchen, there is an old Arata in Khubsugul who keeps a scalp of an almas as a mascot. We are taking steps to obtain it for research. We also mentioned information about skins of almas in one of the Gobi monasteries. We also have personal information from Mongols who in the Barun Hura monastery saw almas skins nailed to the ceiling of the temple. The skins were removed by means of a longitudinal incision in the back, so that the facial skin was intact. The elongated limbs were similar to human feet and hands. The face was framed by dark hair. The informant remembers that the skins were covered with paint and dotted with mystical signs.

In the 1930s a Kazakh primary school teacher and a group of other Kazakhs saw a naked man running on a snow-covered ridge in the Gobi at sunset. He was clearly visible about 100 metres away. He was hairy and running through the snow waving his arms, his reddish hair was glinting in the rays of the setting sun. The teacher wanted to shoot him, but another Kazakh warned him that killing an *albasty* would bring bad luck.

Another report from Dr. Rinchen tells of a man who in 1951 met a naked hairy man in the steppe area and was very scared.

There are a number of stories coming from the Kazakh area, for instance a story of a schoolteacher who had seen an almas. And in 1951 one was seen in Saira Oh. It was a creature of great height, and it scared everybody. In 1948 a hunter found a small creature the size of a 15–16 year old teenager in his trap, but took pity on it and released it.

One of the descriptions is highly detailed. It took place in early May 1938, in the mountains of the upper reaches of the Burhutu Delyon Gol River. One of the participants in this incident, Nagmo, who now works in Ulan Bator, was interviewed separately by Prof. Rinchen and the zoologist Zevegmid, and the Polish correspondent Jerzy Zelensky. Having gone deep into the mountains, a group of hunters found tracks of bare feet in areas of melted snow, and in places where the snow was deep they could establish that whoever made the track was naked, and indisputably male. The hunters then saw a large man crossing the terrain about 100 metres away. He was no less than two metres tall, powerfully built, with a mane of light (possibly grey) hair on the head and short dark hair on the body. Two local Kazakhs called him *Albastyi* - "the master of the mountains", which cannot be touched. They tried to lure him closer with clothing and food, but although he showed curiosity he did not come closer than 100–120 m. Then they tried firing several small calibre gun rounds near his feet, but that just made him stare at the places where the bullets struck the stones at his feet. The local hunters immediately demanded the cessation of firing, and the almas was allowed to get away.

From the same area, which lies in the north-western part of the Mongolian People's Republic, is a case reported by an archaeologist called Dordzhisurunom. This took place in the late 1930s. One day in early spring a hairy naked man ran into the camp and was driven away by the dogs. The women thought it was a man-eater. One man tried to follow the naked hairy man, but was never able to overtake him. He found tracks of the creature's hands in the ground, and it had widely spaced fingers. Spring was bad in that year, and a lot of other Gobi animals had shown up in areas where they had never been seen before.

We have many reports from the Gobi area about the various habits and properties of the almas. It is difficult to ascertain which are reliable and which are mere folk tales. For example, on the one hand, we have a whole series of allegations of attacks by almas and *almasok*, and them dragging away anything that didn't fight back to their lair. Some of the confrontations often happened when people were brewing or distilling.

There is also a detailed story about an old Mongol called Tsedena, who told about a friend who was part of a caravan walking towards Inner Mongolia. Whilst out looking for runaway grazing camels, he was trapped by a saxaul almas. Older people in the caravan said that the almas would not kill the man, but that it would be on guard for a few days. On their way back, the men in the caravan found the cave of the almas in a sandy ravine, and shot a biped at dusk. Its body was covered in hair. They found the man from the caravan in the cave. However, when he returned home, he was different and would not look people in the eye. He died two months later, as if devoured by some strange disease.

Most stories however describe the almas as harmless. Usually it will try to escape from man. One of the Mongolian stories calls the almas or *hong-hara-guresu* ("black man-beast") a good-natured beast, which may even at a meeting give people a hug. But woe to those who would meet him in a moment of rage. Then he will pull a tree from the ground with its roots, and use it as a club.

A most interesting story concerns a *tsagdy* (border guard – until 1929). He lived alone in a small yurt on the border in the Gobi. Once on the hill behind his tent he saw what he thought was a pile of stones. When he got closer he saw a naked hairy creature, which jumped up and ran up the hill. The guard rushed into his tent and watched the creature from there. The almas climbed the hill and disappeared behind it. When it had gone the guard followed it. The tracks looked like big human bare feet. It turned out the almas had built a little shelter from the stones as protection from the wind. Since the guard had scared it, it never came close to the yurt again.

Among other strange habits of almas described in Mongolian stories one can mention: going to the tents to pick up a human child, and putting it down again; sitting close to a dying fire to get warm, but not knowing how to make it light up again; may be attracted to the smell of burning tamarisk tree. In Mongolia the main food of the almas is said to be marmot, which they dig up from their holes.

Prof. G.P. Dementiev has presented us with a number of interesting data. It is important to remember that in the past there existed a number of wall paintings of the wild man. They were particularly fine in the Erdene Zuu monastery, built in the second half of the sixteenth century near Kharkhorin at the ancient capital of the Great Khan Ogedei . Unfortunately, today these frescoes are almost completely destroyed, but could probably be restored. Along with these testimonies, going back to ancient times, Dementiev also noted several oral reports on more recent meetings with almas: in 1946 in the Khovd aimag (province); in the same aimag in 1948 where a male and female were seen; in 1956, a Kazakh hunter named Myrzagali hunting in the western parts of the Mongolian Altai was captured by almas but managed to escape.

Summing up all the available data on Mongolia, Dementiev outlined a number of preliminary conclusions. First of all the many similarities in all the reports should be noted. According to him the wild man has been known since ancient times in many parts of Mongolia, and not only in the mountains, although it should be noted that the entire country is of a significant height. Most of the collected reports on the wild man refer to the Gobi desert, located at the border with China. It is assumed that this creature lives in mountains and deserts, even sandy areas. It is quite difficult to date the origin of the Mongolian stories. Many of them are based on sightings 50–80 years old. We have also heard stories closer to contemporary times, but their authenticity cannot be established with certainty.

According to all these stories, Dementiev described the almas in the following way: They are strong animals with broad shoulders and long arms; their fingers and toes do not have claws, contrary to what Przewalski says. According to the Mongols, that is why the tracks of the almas are very different from bear tracks: they have no claw prints, and the location and proportion of the toes are the same as those of anthropoids. This is also consistent with the data of the British explorers in the Himalayas. The coat is brown or grey (again contrary to what Przewalski said about this), rather sparse, especially on the abdomen; the hair on the animal's head is heavier and darker in colour than the rest of the body. Females have extremely long mammary glands. The overall dimensions of these animals are difficult to specify, but they are about the same as a grown man. Bipedal locomotion is common, but sometimes they move quadrupedally as well. They are mainly nocturnal. They are skittish and suspicious, but not aggressive. Almas seem unsociable. Food is mainly small mammals and vegetation. Almas have not developed the ability to speak. They cannot articulate a single word; they do not use fire or any weapons.

The following message from Dr. Rinchen is also very interesting: “In the 40s, I heard from several soldiers that ... in the area which lies at the Chinese-Mongolian border guards had shot two or three almas. They thought it was the enemy crossing the border, or a spy trying to escape, but it turned out that it was several unfortunate almas who ran into trouble with the border guards”.

Recently the section chief of the Moscow wagon factory, G.P. Kolpashnikov, described his memories of an event that took place in August 1939 in an area of the fighting near the Khalkhyn Gol River. As the then head of the special department of one of the Soviet forces, he was summoned at night to a Mongolian cavalry unit, where something extraordinary had happened: the sentries saw two figures coming down the ridge, made the necessary warnings, and then, assuming that it was Japanese scouts, opened fire and killed both on the spot. They were quite surprised when it turned out they had killed two ape-like creatures. Kolpashnikov arrived early in the morning by armoured car to the scene, and examined the two corpses lying on the ground. He said that he “felt some unease as the dead were not enemies, but some kind of animals or two strange-looking creatures”. Of course – he continues – “at the time I did not know anything about the 'snowman', and I do not claim that these were 'snow people'. But at the same time, I knew that they were not apes. The question was of course - what were they? Only one old local Mongolian said that these were the so-called 'wild men' (or maybe 'mountain men'), that lived in these parts. Because of some old superstition, he was afraid to approach the corpses”. According Kolpashnikov the dead creatures were about the size of a human. Their body was covered with reddish-brown hair. In some places it was thick, and in some places the skin showed through. He could remember the thick hair falling down over their foreheads and eyebrows. The face looked "like a very gross human face". Later a Soviet officer said that because of it being wartime, there was no way of sending the carcasses to a scientific institution.

There is also a completely independent message from M.H. Tairov from the city of Harbin, which must be linked to the previous one. In an appended map areas adjacent to the Greater Khingan Range, delineated by the Hailar, Khalkhyn Gol, Tor Gol, and Chol Rivers, are said to have a population of "hairy men". They are as big as humans, but have great strength and stamina. They are covered with dark hair, but are not bears. They are humanoids, and they walk on two legs.

North-east China
Engineer V. Shirokolobova heard in 1953 that in the Taiyuan area of Manchuria a humanoid creature was killed by hunters somewhere in the highlands. It was a large creature, overgrown with hair, but with smooth face and hands. In 1954 an old experienced Chinese hunter, who had lived all his life in the area, said that a few years ago he had found the tracks of a creature much larger than a person.

Some time before 1914, somewhere deep in the forest of southern Manchuria, N. Baikov and his companion, the hunter Boboshin, saw a tame animal-like man in the hut of a Manchu hunter called Fu-Tsai. How the wild man came to Fu-Tsai they could not say, as the information he gave them was somewhat contradictory. Apparently Fu -Tsai had found the young creature in a wolf den almost twenty years earlier. And then apparently he had found out that it was in fact his nephew, who was born half dead, thrown over the fence and carried away by wolves. This last part is of course impossible. Anyway, the wild boy lived with Fu-Tsai, catching birds and animals, squirrels and grouse, all of which he devoured raw. Baikov describes him as short and stooping, dressed in rags. "His hair was matted like a cap. His face red-brown, like the muzzle of a beast. He had a big mouth, with strong teeth and sharp fangs. When he saw us, he sat down, lowered his long hairy arms to the floor, and moaned like a wild animal. His eyes glowed in the dark like a wolf. He also growled, but then went away to the outer wall, where he lay on the floor, curled up like a dog." *

At night the wild man, called Lan-Jen or Lan-Zhen, would run with the wolves. Baikov and Boboshin went out and saw him squatting on his haunches under a tree, where he would raise his head and howl like a wolf, and then lower it and howl at the ground, just like wolves do. Soon they heard similar howls from the nearby hill. Soon three wolves approached, and Lan-Zhen started crawling towards them. He imitated the wolves perfectly in voice and behaviour. The animals let him take five steps towards them, and then they slowly started running back to the forest. Lan-Zhen stood

* *Translator's note: Here follows a whole series of disjointed sentences, probably notes for future rewriting. I have decided to delete them, as they do not give much meaning, and contain very little information, apart from comparing the wild man to a wolf.*

up, and started walking, and then running after them, and disappeared into the forest. Boboshin wrote: "God knows what he does at night in the woods. Nobody knows, not even Fu-Tsai, and he would not tell us, even if he knew." In the morning Lan-Zhen came back. For food Fu-Tsai gave him a skinned carcass of a squirrel. "He grabbed it with both hands, raised it to his mouth, and began to eat the head, and crush the bones with his strong teeth. He growled with pleasure while he tore the meat apart with his hands and his front incisors...After eating, Lan-Zhen went to the water boiler, took out a ladle-full, and drank it in one gulp..."

It is difficult to ascertain the value of this story. If it is accurate, it may be a case of taming and training a relict hominid (the boy raised by wolves is highly unlikely). We do already have a number of reports of taming these creatures, and stories about their amazing abilities as hunters. In folklore they are known as "beastmasters", "huntmaster" and so on. According to Baikov, Fu-Tsai would communicate with this creature and guide it. That does not disqualify the story as such, although it may be a case of a subjective illusion experienced by the observer. We do after all often speak to the family dog, and although its brain is far less developed than the brain of a hominid, a dog can distinguish and understand very complex verbal commands.

We can now finish our review in the southern Shaanxi province in the People's Republic of China, near the Qinling Shan mountain range. This is in a sense a continuation of the Nan Shan range, and here we already have a number of reports of the wild man. The Qinling Shan mountains are also closely linked with the Sichuan Alps, adjacent to eastern Tibet, and the mountains in the western part of the Yunnan Province.

A native of Shaanxi Province, now chairman of the government of this province, Chao Shou-shan, has said that about forty years ago he personally saw *Zhen-shu* ("human-bear", "wild man") in the Qinling Shan mountains, but not close. He was on top of one mountain, and the Zhen-shu on the other. On another occasion, a hunter had killed a Zhen-shu and removed its skin. In general, residents of the Qinling Shan are well aware of the existence of this creature, and know it lives a solitary life, that it cannot speak, but laughs and cries. One story says that a hunter one day came very close to a wild man, and that the wild man put his hands on the hunter's shoulders and started laughing. They also say that in recent times a Tibetan woman had been kidnapped by these creatures, and that she was found two years later by a Chinese military patrol.

There can be no doubt that when the famous Chinese historian Prof. Hou Wai-lu told us about human-like wild creatures living in the Qinling Shan Mountains, and that they had been known to the people in the area since ancient times, he was talking about the same beings. There can be no doubt of their existence, he said. Before the people's democratic revolution, the residents of the southern regions of Shaanxi Province did not consider them human beings, and hunted them. Any survivors were captured and tamed. They were used as labour and could perform simple tasks. Professor Hou Wai-lu considered this practice inhumane, comparing the prisoners to captured slaves. He thinks they are descendants of the Western Zhou tribe (tenth–eighth centuries BC), who became marginalised and returned to a feral state in the surrounding mountains. However, when studying the specific details of these "wild men" of the Qinling Shan mountains, he had to concede that there were structural differences between them and humans, but he also felt that they could to a large extent be explained as subjective experiences in the observers and a form of selection. So according to Prof. Hou Wai-lu, the strength and the hairiness of the wild men can be explained by the fact that they live naked in the cold highlands. The physical evidence is much more important. They have no clothing or shelter, they live on raw meat and wild fruits, they use no tools, and they have no speech, no language.

Prof. Hou Wai-lu is in no doubt that a population of wild men can still be found today in the mountains of Qinling Shan. In 1954 he was able to study the traditional methods of hunting for them in a mountain village. The people would lay out pieces of red cloth in several places on the slope of the mountain. This would attract the curious creatures. Prof. Hou Wai-lu was able to study one of these captured creatures. It had been caught a few years before. He had been tamed and learned to perform simple work. The most interesting observation was, that although the wild man could not speak, he learned to play a few simple Chinese games.

CHAPTER 7 • SOUTH OF THE GREAT HIMALAYAN WATERSHED

In Chapter 4 we reviewed the many studies done in the 1960s in the lands south of China – chiefly Nepal and Sikkim. The research in this area has not only given us a wealth of descriptive material, but also a small amount of real data, that has been described above or will be the subject of special analysis in a later chapter. This data is divided into 3 groups:

1) a series of photos, casts, sketches and descriptions of footprints in the snow;
2) photos and hair samples;
3) a series of photographs, samples of dried skin and muscle tissue and hair taken from the mummified hand in the Pangboche monastery.

To this list we can add indications of other possible material remains of the yeti, which may still be available for study. Professor Y.N. Roerich reported that a Tibetan living in Kalimptonge kept a piece of skin from a "wild man". It was supposedly carved from the back of one such creature. The owner showed the skin to Professor Roerich. According to the latter it was a fairly thick piece of skin, brown and with fairly long hairs. It was clearly different from bear and langur. In addition, Roerich heard from Sherpas and Indian zoologists that a monastery in the south of the country had among other things a scalp and a skull of a Bigfoot.

When the Director of the State Museum of Nepal, Professor Chandra Man, visited Moscow in June 1958 we asked him questions about the Bigfoot, which he had never actually studied. He said that the Nepalese State Museum had a picture of the mummified corpse of a baby "snowman". The photo was taken when the mummy was deposited in the museum for a few days by its private owner, a person residing in Kathmandu. He had gotten the mummy from "the other side of the Himalayas", i.e. from Tibet. It was stored in a special container covered with sugar to protect it. The corpse was from a newborn child, a male. In its shrivelled state it has a total length of 20–23 cm. The skin of the mummy is of a dark brown colour, with no hair. The eyes are very small and sunken. The hands are disproportionately long compared with the figure of the modern man. The overall impression according to Prof. Chandra Man was not of a monkey, nor a modern man, but rather a prehistoric one. In 1960 this case was taken up by Professor Corrado Gini (Italy) who visited the National Museum in Kathmandu specifically to study the "baby snowman". He was able to show that it was in fact a fake, and that it could not possibly have a real biological origin.

The next incident is reported by two Americans, George Moore and George Brooks, and it was first published in a recent book by Ivan Sanderson. The incident took place in 1953 during a descent from a pass on the way to Kathmandu. They were in a forest in thick fog, and were suddenly scared by strange voices and shouts nearby. The sounds were coming from several sides, and the travellers realised they were surrounded. According to George Moore, “a disgusting person suddenly stepped out of the mist, and I will never forget it”. The skin was grey, the eyebrows black, the mouth stretched almost from ear to ear. It had long yellow teeth and prominent upper canines, the eyes were small and sunken. The whole body was about 150 cm in height, the posture slouching, legs were hairy, hands looked dark, body was muscular but grey and dirty. A moment later six or seven other individuals stepped out of the fog, one with a child hanging around its neck. The travellers fired shots in the air. This made the creatures stop, and let the travellers escape.

One of the few who has taken the testimony of the Sherpas seriously is Izzard, who has written extensively on the material he collected from the expedition in 1954. “Based on my rather extensive experience with the peoples of Asia, I would argue, and I'm sure that every traveller in the Himalayas will certainly agree with me, that the Sherpas are as intelligent as we are, have the same sense of logic and are perhaps even more sober. Many of these people spend most of their life in the mountains from early childhood, looking after their herds, roaming the plains and the distant slopes; if we take into account the rather limited emotional nature of the people, it should be clear that to ignore the information that you can get from them would be as stupid and harmful as to doubt the stories of a forester from the north of Scotland about a noble stag or a Sussex shepherd of the habits of the hare ... the yeti was an indisputable fact, many people have seen it and consider it a common animal, along with wolves, leopards, wild goats and other animals living in these mountains.” “Not once did I have the slightest suspicion that my interlocutors said what I wanted to hear, or invented an interesting story especially for me.” “We questioned the inhabitants of ten villages in different parts of the country. Their stories have been logical and devoid of any contradictions ... When trying to recreate the appearance of the animal based on the stories, we discarded all second or third-hand information, and used only descriptions from people who claimed to have seen it themselves. No matter where, when and under what circumstances Sherpas had seen the yeti, they gave exactly the same description... We found that the local residents tell very plausible stories about the unknown animal, that there is a large unexplored area where it could live, and that it lives on plants and animals. One of the main proofs of its existence is mysterious footprints which have been seen in the snow in several regions of the Himalayas. Unfortunately, that gave a completely false picture of the animal as an inhabitant of the zone of perpetual snow. All Sherpas agree, that the animal lives among the rocks. This is not a desert zone, but alpine country with dwarf shrubs, flowers and herbs. It is populated by a considerable number of birds and mammals. It is here and only here one will be able to find the yeti according to the Sherpas. Food is abundant. A significant portion of the area is teeming with pikas, small animals that look like guinea pigs. The Sherpas and the Tibetans believe that pikas are the main food of the yeti, supplemented by insects (in the summer), and probably the roots, shoots and leaves of alpine plants."

Kashmir

The mountain areas of Nepal and Sikkim, the northern part of the United Provinces of India and the states of Jammu and Kashmir have not yet been the subject of systematic information gathering. However, a few records of *vanmanas* (wild hairy man) are known from these areas, especially Kashmir. Several peasants who live in the valleys of the area have claimed that they have seen these strange animals and their tracks. One farmer who said he had seen a female with cubs described their long greyish hair. Another story talks of a man who saw a hairy creature eating rice from a pot. When it was discovered, it ran off. One report tells of a vanmanas killed near Gangabala in Kashmir, but a detailed description of the body could not be obtained. In the Narang area of Kashmir there are also stories about vanmanas kidnapping people.

The most detailed information about Kashmir was collected by the social activist Mirabehn (Madeleine Slade), an English follower of Mahatma Gandhi, and given to Odette Tchernine in 1959. She said that in 1951–52, a "Bigfoot" was seen at a distance, in an area about 500 miles west-north-west of Mount Everest. Farmers in Garwhal also gave her information about tracks of the yeti found at a height of 4000–7500 feet, i.e. well below the snowline. One old man who was working near his barn saw a "Bigfoot" on the mountainside in the moonlight. This creature was very tall, "like a giant hairy man with broad shoulders, as if he was wearing a shaggy coat" - these are the actual words of the observer. The next day he found footprints. Another man claims to have seen a being, standing up straight, while another one, smaller in size, was running around it. Mirabehn collected a number of reports in 1954–1955 as well. One was from a man who was tending cattle around some building at a height of about 6000 feet. From here he visually observed a "snowman" at a height of 4–5000 feet. When questioned the local farmers were absolutely convinced of the reality of the yeti, and considered it a rare animal.

Mirabehn also showed the peasants photos of tracks found by Eric Shipton in the Himalayas in 1961. They all said that they had found the same tracks, not only in the snow, but also in the fields. One of the farmers said that he had seen similar marks on the ash heap. All these tracks have been seen in the Garwhal area at heights of around 4000 feet. One man, who was a soldier in the Indian army during the Burma campaign, told Mirabehn: "I saw the 'Bigfoot' once. I looked at the side of the mountain, and noticed a creature with long red hair, who was relaxing in the sun. I got scared and threw a stone at him, and then the 'man' ran away." In general the people of these regions speak of the yeti as a special kind of people. Mirabehn also talked to Kashmiri nomads who claimed to have seen both female and male yeti. In the two years before she resumed the quest for the yeti, a *vanmanas* abducted a woman from a local tribe. The shepherds chased him through a gorge with a cave at the end. The entrance was closed with a large stone. With great difficulty, they managed to roll away the stone from the entrance, corner and kill the "hairy man". The woman escaped unharmed. When Mirabehn heard about the story she tried to talk to the witnesses, but they refused to talk to her because "she had something to do with the government", and they were afraid the shepherds would be accused of killing a man. Another story is about a group of men digging a ditch for water two miles from a camp in the alpine meadows where they were tending their herds. One day at a distance of about one mile from the corrals, the man who watched over the channel came face to face with the vanmanas. It was looking at the newly dug earth, as if surprised by this intrusion into its personal possession. The shepherd was afraid and, with his fingers in his mouth, whistled loudly, calling on the help of the others. Hearing the sound, the vanmanas looked back, jumped over the freshly dug ditch and disappeared into the woods. When the people were interviewed, they described the face of the creature as a human face. One of the Kashmiri men compared it to a photo of a businessman in a newspaper. The people were certain that vanmanas was not a bear or a big monkey, and it did not walk on its knuckles.

"The only way to establish the truth about our snowman is to approach the issue with caution" – says Mirabehn – "and not to make so much fuss and noise about it. You will only succeed with infinite patience. We should not kill or capture the beast. We should hunt it with a camera. The best way to move forward in the search would be the organisation of small groups of local people. It should be locals. The yeti knows their smell. They should be trained in handling cameras, and then perhaps after waiting patiently under cover of branches for many nights, they may get a result."

Nepal and Sikkim

A lot of material has been collected among mountaineers in Nepal. Two features stand out, when you look at the material. There is almost no mention of young or of females. But there are descriptions of what seems to be two or three types of creatures with different names and different looks, from giants almost three metres tall, which only very few people have seen with their own eyes, to very small creatures not more than 1.2 metres tall. It is possible that they are part of a migration across the Himalayas, and that only adult males and half-grown individuals take part in this. We can also assume that neither Nepal nor Sikkim is part of their breeding grounds.

In 1953 a series of sightings around the Tengboche monastery was recorded. Several yeti were seen close to the walls of the monastery. But please note that this is not the same observation as the yeti seen by a group of Sherpas at the same location. The various sightings were examined and shown to match closely. Here is one example. In the winter, when the ground was covered with snow, a yeti appeared from the surrounding thicket. The beast moved in huge leaps, sometimes only on its hind legs, sometimes on all fours. It was about 1.5 metres tall and covered with grey hairs. It stopped to scratch itself, then picked up a lump of snow, played with it and growled. People at the monastery decided they had to chase the animal away. (Stonor adds, that the monks thought the beast was also watching them.) The monks started to blow on long horns and conch horns, and that made the yeti trot away into the bushes.

Tenzing Norgay, the famous conqueror of Everest, also had information on the yeti. He said he had heard about the yeti from childhood. People in his native Khumbu talk about them a lot. As a boy he sometimes found excrement of an unknown animal on rocky slopes and glaciers. It contained bones of small rodents. He was sure they were from the yeti. He saw tracks of the yeti twice: in 1946 on the Zemu Glacier, and in 1952 at the foot of Everest. He also told some very interesting stories about his father: "The first time he met this strange animal was on the Barun River near Mount Makalu, near Tse-chu where I was born. He came upon him suddenly, and was so close he saw him quite clearly. He said the yeti was like a monkey. The only difference being that he had very deep-lying eyes and the head tapered at the top. The body of the animal was covered in greyish hair, and it grew in a very remarkable way. From the waist up, it grew up. From the waist down, it grew downwards. It was a female with flabby breasts, and a height of about 1.2 m. The animal moved only on its hind legs. Father was very frightened, but so was the yeti. The animal suddenly turned sharply and began to climb up the steep slope, uttering a shrill whistle, and then it disappeared. Following that experience, father was expecting trouble for a long time. Many people claimed that you would die if you had seen a yeti. He did not die, but he was ill for almost a year after the incident."

Tenzing's father met the yeti for the second time in 1935, this time in the Nangpa La Pass in Rongbuk. He happened to spend the night alone in the intermediate camp of the climbing expedition. At dawn he heard a shrill whistle, and when he looked out of the tent he saw a yeti walking on the glacier going from south to north. Tenzing's father waited and watched until the yeti was out of sight. Tenzing also told about one of the porters of the Everest expedition. He had bumped into a yeti at a place where a few had already seen its tracks. "He said it was about one and a half metres tall, was covered with thick brown fur, and walked on its hind legs. The head was pointed at the top, had high cheekbones and powerful jaws. The yeti bared its teeth in a threat, and stared fixedly at the porter, as if it was going to attack him. Then it suddenly hissed, turned and ran away." "The words of the porter only confirm my impression," said Tenzing, "that the yeti is a big mountain monkey." "Although I cannot prove it," says Tenzing elsewhere, "I am sure it lives here. I think it is a beast, not a human being. It only leaves its lair at night, and it feeds on plants and small animals that live in the alpine meadows. It is probably an unknown species of monkey."

The information collected by Professor Y.N. Roerich in the same areas is particularly interesting. "In my time in the Himalayas, and I have lived there since 1923, I have had to ask the Tibetans and the Sherpas and other people in the region about the 'snow people'. They all agree that they are real. The Sherpas generally think that the 'snow people' live on the southern slopes at a height of 5–6000 feet. They call it the *de-ti*, whereas in Tibet the animal is called *mi-gyo* ('wild man'). According to the Sherpas it is a rather large monkey species. It is upright, and about 150 cm tall, covered with fairly thick, but not long, brown hair. It lives apparently in mountainous conifer forests with rhododendrons. It eats fruits, berries and roots, and apparently never attacks humans or large animals. It has been seen, but only at considerable distances, walking like a man, and not on all fours like a monkey. The creature often gives a very distinct whistle. According to a former British commercial agent in Gyangtse in the Chumbi Valley, he heard the distinctive sound near the Dronkyala high ridge, and when he asked the guides what it was, they all said 'wild people'." It is very clear to Roerich that the slopes of the Himalaya Range are home to a creature, perhaps a large ape, which is unknown to science. It is not just folklore, and the people who live there are certain of their existence. The Sherpas and other tribes who live in the area know the bear, but they still insist that there are wild men or apes in the area.

The anthropologist Professor Nebesky-Wojkowitz has concluded that "It is significant that the statements of the Tibetans, Sherpas and Lepchas about the appearance of the Bigfoot coincide." It is 2.1–2.3 m when standing upright on its hind legs. It is strong, has dark brown hair on the body, long arms, and an ovoid monkey-like head which is pointed at the top. The face is covered with sparse reddish hair. Despite its immense power, the Himalayan people consider it a harmless creature, that only attacks humans if it is wounded. According to a local hunter, it is wrong to call it a "snowman", because it is not a man, but an animal, and it does not live in the snow. It prefers the band of high-growing dense forests in the Himalayas. During the day the animal sleeps in a shelter. At dusk it starts to move. You can hear it approaching from the sound of breaking branches and the whistling cry. In the dense forest, it moves on all fours or jumps from tree to tree, but in open spaces mainly on the hind limbs, but with an uncertain, swaying gait. It actually does not like to move in the snow, but the local people say it is because it likes a special type of moss that contains salt, and it is only to be found on the rocks in moraine fields. It is only when searching for this moss that the creature will move through the snow. When it has fulfilled its need for salt, it will return to the forest.

The material gathered by Stonor has given similar results. When he talked to the Sherpas in Namche Bazaar, they all said the *ye-ti* was a beast, and not a man. It had a face like a monkey and the body was covered with hair. The hair on the upper part of the body grows upwards, but from the waist down it grows downwards. According to the Sherpas the yeti lives in the higher mountains over their villages, but it descends to lower altitudes in winter. This animal lives only in the vast area with loose boulders in the high alpine zone where there is no woody vegetation, but still below the permanent snowline. At times it will approach people, and it can be seen and heard near the villages. It has the same general build as a man and no tail. The fur is reddish brown and black, but with a slightly lighter colour on the chest. According to most reports, the hair on the head and the middle torso is especially long. The head is pointed at the back. The face is naked. It would make a loud, extended and high-pitched scream. It can be heard most often in the afternoon or early evening. During the cold winter months, the yeti comes down closer to human habitation. As suggested by the Sherpas, snow and storm forces the animals down the mountain. It usually walks on two legs, but when in a hurry or trying to get out of deep snow it will use all four legs. Many people believe the yeti eats pikas as well as other small animals and insects. The animal is very careful and clever, and Sherpas treat him with great respect. They are afraid of it, although there are no known cases of it having assaulted humans. The following are some of the most interesting individual records collected by Stonor:

Tenzing Lakhpa, a shepherd, 30 years old, talked about an incident that happened three years ago, in early spring. While searching for a straying yak, he said "I climbed up on a group of rocks overlooking the pasture. There I heard the sound of screeching. Coming closer to find out what was happening, I found some freshly killed pika on the ground. A little further on, at a distance of 30 paces, I saw the creature. I recognised the *de-ti*. He sat upright on a rock like a man, but with his back to me. He was of the height and build of a 12-year old child. His head was pointed, and his back was covered with reddish-brown hair. I was very scared when I realised what was sitting in front of me, and quickly crawled out of sight before he could sense my presence. "

One Sherpa told Stonor that two years ago, on a spring day, when he was working in a birch forest in the mountains, he suddenly heard someone coming down from the alpine zone above the forest, making their way through the undergrowth. He crouched down with a club in his hands, but the unknown beast approached him until it was only a few steps away. Due to the dense thickets of rhododendron, he could only see that it was of medium size, with dark hair, and looking like a person. The creature was so close to the frightened Sherpa that he could hear it breathing. It sounded like a man breathing, not at all like the breathing of an animal such as a leopard or a bear. A few minutes later the unknown being disappeared back in the same direction.

A Sherpa named Nimah told how, when he spent the night with a group of companions in the mountains in an abandoned shepherd's hut, they had heard the approaching cries of a yeti. It was so close the hut was vibrating with the sound. At dawn they found fresh tracks in a thin layer of snow. It was of a biped, and looked similar to a man with small feet. The tracks led directly to the hut, and then away again and disappeared in the distance.

The mother of a lama saw a yeti at a short distance; the creature looked very similar to the animals described in other stories. A lama from the Tengboche monastery claimed that he had seen a yeti twice in his life. The last time was three years earlier, when a yeti came down from the high mountain during bad weather. The monks at the monastery had to make a lot of noise to make it go away. A villager named Pangboche Dakhu said that four years ago, while he was out looking for a straying yak among the rocks, he saw a hairy beast hiding behind a rock. The beast got up and slowly walked towards the Sherpa on its hind legs. The beast was very heavy and menacing. It tore whole bunches of grass from the ground. The colour of the fur looked like a goat. It was black with a tan tint and a lighter shade on the abdomen.

The hair was quite long. The animal resembled a stocky person. The threatening persistence of the animal forced the shepherd to flee. In the mean time, two brothers nearby had heard cries of a yeti close to some huts. The next morning they found tracks like human feet on the ground. They were about a foot long.

In March four years ago a resident of Namche named Mingma, a shepherd, heard the cry of the beast in the pasture above the village of Pangboche. He then saw the beast coming towards him. It was hairy and walking on its hind legs. Mingma ran to a hut and barricaded the door, but watched the yeti through a gap in the wood. It was a squat stocky animal, built like a small person and covered in reddish-black fur. The hairs on the legs were longer. The elongated head had a topknot on the crown. There was no fur on the face except for a small amount of brown hair growing along the cheeks. The muzzle was flatter than a monkey, but less flat than a human face. The nose was flat, there was no tail. The yeti walked slightly bent over, with hands hanging at its sides. The hands looked bigger and stronger than human hands. After standing still for a bit, the yeti walked in front of the hut with long strides. It noticed Mingma peering at it, snarled and bared its teeth at him. The teeth were much bigger than human teeth. Mingma finally drove away the yeti with burning branches from the fireplace. When Mingma later was presented with a drawing of the alleged appearance of the yeti he said the creature he saw was different. It looked more like a monkey, and the hairs on the arms and legs were too long.

Four Sherpas going to summer pasture in the Bhote Kosi Valley saw a yeti perched on a rock 50 metres below the trail. It was brown-black, and looked like a small man with a pointed head. At the sight of the people, the yeti jumped off the cliff and ran away on all fours. In the Phorche village two years ago in the middle of a very cold winter, the peasants one night heard the cry of the yeti, and for two or three days they would see tracks in the snow in the shape of human feet all around their yaks. There were also lines around the tracks, perhaps because there were hairs growing on its legs. The villagers tried to follow the tracks of the yeti. It almost always walked on two legs, but sometimes it would go down on all fours in difficult places. When reaching a flat stone, the yeti sat down on it. There was a print of its backside on the stone, and next to it two dents where it had put its hands. Below on the ground there were two clear parallel footprints. They also found a little tuft of reddish-brown hairs in a crevice in the rocks. Their hairs were more stiff and bristly than the hair of other animals. They did not bring the hair. The peasant believes it brings misfortune.

Sherpa Ang Tsering said that once in the late autumn, he and his wife were collecting medicinal herbs among the rocks at the edge of a forest area, when they scared a yeti. The animal jumped out of a small hole near them, and galloped away on all fours among the rocks. It is very different from all other animals. At the end of the day they also heard the well-known scream of the yeti. One winter after a heavy snowfall shepherds from the village of Khumjung scared a yeti that was standing a hundred yards in front of them. It was a humanoid creature with hair like goat's wool. When it saw the people, the yeti let out a shrill scream and ran up the hill, all the time holding on to its hind legs. It escaped, but did not move very quickly. A Sherpa named Phorchen Da said that he and his friends had scared another yeti in the Longmoche mountains twenty years before. The beast ran away on its hind legs. It was smaller than a person, and with the colour of a musk deer.

The same Phorchen Da has told an interesting story about meeting a yeti nine or ten years ago, at the edge of the forest in the Bhote Kosi valley. He was searching for bee's nests, when he saw a brown beast coming down through the trees. At first he thought it was a musk deer. However, when the animal came closer, it became clear that it was walking on its hind legs. Phorchen Da hid behind some boulders and kept watching the animal. It walked slowly on its hind legs, just like a man taking big steps. At the more difficult places, he did go down on all fours. It was the size of a boy. It had shaggy reddish-brown and black hair, but it was lighter on the chest. The hair was not very long, but looked like the fur of a goat. The head was elongated and pointed, and the face looked like a monkey, as it was hairless and brown. Eventually the animal climbed on a rock a hundred yards from the observer, and just sat there on its haunches. Phorchen Da decided to drive the creature away, and started beating pieces of wood together. The yeti jumped off the rocks and moved quickly away on all fours. The next morning Phorchen Da and a friend found tracks of the animal at the rocks. They looked like small human footprints with very clear lines.

One story that deserves more attention was told by one of the senior monks in Pangboche. Recently he climbed up on a deserted rocky slope a two-hour walk from the monastery, and found a large nest of interwoven twigs of fresh juniper. It was like a big bed, and big enough to accommodate a human. The monk had heard stories about yeti nests, and was sure he had found one. Another old man found a nest among the rocks in a deserted area. It smelt very unpleasant. Stonor writes that he interviewed Sherpas about these nests, and although they are very rarely found, they always emanate a terrible stench.

In the winter of 1957–1958, two representatives from the Sherpa village of Kullu appealed to the Nepalese government for protection against a yeti. It had climbed into a mill, started eating the miller's meal and smashed his grinding stone. This case is not completely exceptional. In early 1958 residents of the village of Tarka, three days from the capital city of Nepal, reported a similar visit by a yeti to their mill. As one of the farmers was going to the mill in the morning, he saw that the door was open. He peered inside, "but I did not see a man. It was a huge humanoid creature, all covered with thick long hair. It was eating all the flour and grain it could find. The creature was about 3 metres tall, and the arms were long enough to touch the knees. It was intent on searching for food, it did not even notice me. I have heard many stories about these creatures, and know it brings bad luck to meet one, but I still decided to look at it properly. It had a flat bald face with lots of wrinkles like a monkey. The head was high and conical, and the body was covered with hair. It had no tail. The nails were like claws, and it kept roaring. The whole body of the creature was white with flour. The yeti had not noticed me, so I walked carefully away, and ran to the village to get others to help. But the yeti heard me, and started climbing toward the snow-covered mountain peak. The villagers who heard my screams saw him running away." Three villagers confirmed the story.

The following testimony was written by Peter Byrne and Gerald Russell in early 1958 concerning an incident on the road to the Barun Valley. A yeti had been seen just two days before, and the two eyewitnesses, a boy and his older sister, were interviewed separately. They were grazing their herd of yaks at an altitude of 4500 m. Sometime in the afternoon, a large upright animal came out of the bushes and approached them to a distance of about 30 m. The boy said the animal was about 3 m in height. According to both, the creature was covered with black and brown fur, except for the waist, where the fur was whitish. The head was high and conical. Both eyewitnesses were shown various pictures of other animals as well as reconstructions of the yeti and of prehistoric man. Both said the last two were most accurate. They also noted some similarities to a photo of an orangutan, but rejected pictures of black bears, monkeys and other Indian animals. Perhaps the most interesting aspect of the case is the fact that, after having run away from fear, the two shepherds returned to their animals. The yeti had disappeared, but one of the yaks was lying on the ground with a broken neck, as if from a blow.

This case may not be trustworthy, but it must be compared with the words of Tom Slick: "along the way we often heard stories of the yeti killing yaks. It apparently breaks the neck of a yak or drops it off a cliff or a rock. It does not happen often, but usually a few times every year. But the yeti, which is clearly not a carnivore, does not eat the yaks… It is generally believed that the diet of the animal consist of bamboo shoots, of which there are plenty. Perhaps they also eat other roots, different types of insects, small rodents, and perhaps an edible fern, which is very common in the area. But we also heard about some cases, three with the victim's name and the name of the village, where, during the past three years, the yeti have killed people. Allegedly, the eyeballs, fingers and toes and testicles of the victims had been eaten." The latter seems strange, but how much do we actually know about the relationship between the local people and the "snowman"? Stonor for example described a funeral of a poor Tibetan woman from Namche. Instead of burning her, her body was carried from the village up into the mountains and left among the rocks.

Finally we will take a look at some records from the Hillary expedition to Nepal in 1960. Truth be told, there are some contradictions between these records and the writings of Hillary and his literary agent Desmond Doig. On the one hand, Sir Edmund said that the expedition did not meet any Sherpa who claimed to have seen a yeti. On the other hand he has also written for example: "a tough and highly experienced Sherpa told me with absolute certainty that he had seen a 'yeti' and had watched it for some time." We also find a few interesting records in the correspondence of Doig. Soon after his arrival among the Sherpas in September 1960, Doig wrote: "This morning I met two people who saw the yeti close enough and for enough time to give a detailed description. Twenty years ago, the village of Beding two hundred kilometres from Kathmandu was covered by a large avalanche. Some of the people were buried alive in their stone huts. It was in that year, one of the residents remembered, that the yeti made themselves felt very strongly. They made a terrible whistling in the cold winter nights, and every morning their tracks could be found in the fresh snow." The head lama of the local small monastery told Doig that he was certain the yeti were looking for the bodies of the victims who were killed by the avalanche.

Just two winters ago, a pair of "snow people" came up to the monastery at the time of the evening prayer. It was already dark and snowing. They circled the building and even tried to break through a window. The monks were very frightened (there were only five people in the monastery) and started ringing the sacred bell as much as they could. That made the yeti run away howling. Their voices were very similar to human voices, and they sounded full of sorrow. The abbot of the monastery told about another event, that took place several years before. A Nepali official with servants came to Beding in the winter to hunt deer and pheasants. "The group had arrived and was about to go to bed, when they heard the noise of several large animals circling the building. Suddenly a large head overgrown with hair appeared at the window. Nobody dared to move to grab a gun. The yeti howled with rage. The lama heard the noise of the animal, and started to blow the big horn. The frightened animal fled, making a terrible noise." Another lama told the story about how one night, when he was asleep in his cell, he was awakened by a terrible noise and smell. To his utter dismay, he saw a yeti trying to get inside. "It was a real *mi-gyo* covered with black hair. It had big eyes and was about 1.5 m tall." The lama started saying prayers that could save him, and that drove the beast away. The next morning they found traces of very large human-like feet in the snow.

In another correspondence, Doig writes the following: "... A few days before leaving for the expedition in Kathmandu, I questioned a Nepalese, who claimed that he had seen a yeti. This man – Gary Badahur Vhayang – is from a small town, situated at a considerable height in the extreme north-west of Nepal. The city is governed by a Rajah. It borders with the small Tibetan 'state' of Tuglakot, which also has its Rajah. Its capital is a village consisting of shacks built of stone and wood. Between these two states, according to Gary Badahur, lies a valley, which belongs to neither Tibet or Nepal, and which the natives call Mahadip. That's where Gary Badahur and two of his companions saw two yeti. These 'Himalayan people' were on the other side of the stream at a distance less than 200 metres. They were so similar to humans that he shouted to them: 'Where are you going? Hunting?' The two 'human beings', one of whom seemed to be an adult, and the other a child, stopped for a moment, but then fled from the valley in long jumps, and started climbing the steep cliff with no effort." The more information Doig gathered, the less mythological the animal seemed to become. He was told the yeti only appears in the winter, at a time when only very few people oversee the yaks. "They showed me which side of the village the yeti usually visits, where they find a watering place, and which direction they travel. They are mainly associated with the river. They usually wander along the river in search of something, but what? We found no traces of frogs or fish. But it is possible that in midwinter, when everything is frozen, many animals and birds will come to the river." Doig realised that not only local people and European visitors tell these stories, also resident Europeans like the British Agent in Kalimpong, David MacDonald. He is quite certain the creature exists. MacDonald's three daughters travelled from Tibet to India for several years, as they went to school there. On one trip, when crossing the Chumbi Valley and Sikkim, they heard the strange screams, and their porters said it was yeti. The Sherpas said the screams were so terrifying they could paralyse a person.

There is also a letter from a certain Mr E. Hillary Hoof. "I spent almost all my life in these mountains (Himalayas), first at a desk at school, and then for 33 years as an employee of the North-East India railways. In 1925 I was a member of the railway society of hunters and anglers. I often went hunting in the car, especially in the dense Serayu forests. My Nepalese driver, Hurka Badahur, knew the road and was a brave man. One night, when we returned from Kalimronga, which is 65 km from Darjeeling, the car was going silently down the slight slope of the mountain, when we suddenly found ourselves face to face with the creature, standing upright, blinded by the headlights of the machine, about 20 metres from us. I was preparing to shoot, when Hurka Badahur intervenes and tells me in Nepalese: 'Sahib, do not shoot! This is the man!'. I vividly remember the man, or rather, the red brown dwarf. It was no more than 1 metre in height. We expected him to run away down the road, but he leapt across and ran down the side of a hill in the forest."

Here is another of Doig's entries, this time from the Royal Palace in Gangtok. "The Crown Prince of Sikkim, Palden Tondup Manguyal, who is a very intelligent and educated man, suddenly turned to me and asked, 'Do you want to talk to someone who knows a lot about the yeti? Here he is, ask him.' It was one of the prince's personal guards. He did not give the impression of a person suffering from hallucinations. As it turned out, a few years ago he and his friend were attacked by a yeti, just 20 km from the capital. They were out hunting deer in the maze of peaks and ridges of granite rocks that extends from the border of Sikkim to Tibet, above the forests of rhododendrons. According to the ancient tradition of Sikkim, every time they killed an animal they were supposed to immediately use the hooves and insides of the animal as a sacrifice to the patron spirit of hunting: the yeti. However, they decided to continue the hunt, the sun was sinking toward the west, and they would postpone the ceremony till just before dark... As soon as they began the ritual, they heard the noise of rolling stones and footsteps, and felt the horrible smell of the yeti. Then the attack began. Stones and branches were flying through the air, thrown with surprising accuracy. At the same time invisible beasts were howling and whistling. It seemed like there were hundreds of them. The people ran and left the deer. They ran without stopping all the way to Gangtok."

According to Stonor, the only serious argument against the existence of Bigfoot, based on the stories of the Sherpas, is that none of the narrators could find more than one animal at a time. However, the Sherpas themselves explain this by saying that the yeti are very rare, and constantly roam the vast territory of Nepal and Tibet, and try not to be seen. And maybe they were all single males? Maybe the real refuge of the smaller yeti is somewhere to the north, east or west? In spring and summer, when it is easier to get food, the yeti may go there and live in groups. In autumn, when the pikas are hiding in the rocks, it is more difficult to get food, and then the yeti starts to wander through the mountains.

Izzard writes that: "The Sherpas suggest that the females and young animals may live in remote caves or shelter among the rocks and stones, and that the males tend to wander around on their own. I did speak with people, who claim they have seen tracks of several yeti crossing the same area."

The Sherpas think that the Nepalese southern slopes of the Himalayas are home to single individuals in autumn and winter. In the spring, when there are large edible insects and caterpillars to be found, and the pikas emerge from their shelters, it is much easier to find food, and the yeti leave the Sherpa country to the north, and maybe also to the east or west.

Many of the participants in the expeditions to the mountains of Nepal think that the stories of the Sherpas are about two or maybe even three different creatures, not just one. The first type of yeti is the *shu-te* (or *zu-cho*). It is very large, 2 m, has 40 cm long black or red hair and is a vegetarian. In winter it disappears for long periods. It can be dangerous to humans, but only very rarely. The second type is the *mi-gyo*. It is 90–120 cm tall, with a lighter coat, but can also be red, black or grey. This one is more aggressive towards humans and animals. Both of these types walk upright, but sometimes use four limbs. They are very similar to humans, and can make a high whistling sound. Both emit a horrible odour. The third type is *those-ima*. Its relationship to the yeti is less clear. It is 45 to 60 cm tall, lives at lower altitudes in the mountains, and seems to be able to build simple dwellings. But why should it be considered three species? It could easily be age-related, or differences related to biology and habitat. After the 1958 expedition Gerald Russell realised that the most numerous type is the smaller one. There are fewer of medium size, and the large ones are the rarest. Based on these numbers, Russell thinks that the number of "wild men" in the Himalaya region is about 4000. That is not many in relation to the size of the Himalayan area.

According to Stonor there are no mandatory rituals or spells to perform for individuals who have seen or heard a yeti, but it is fairly common to do it anyway. The most pious Sherpas definitely recommend it as not something to neglect. Stonor also warned that monks are much less willing to disclose information about the yeti. Lay people generally believe that meeting with a yeti is a bad omen, and entails illness, death and misery, so they hurry to the lama, who makes the proper ritual and turns the bad influence away. The meetings between the Sherpas and the yeti are never associated with any kind of rituals or spells. It is always accidents happening during their everyday lives. Thus you can make a clear distinction. Among the Buddhist clergy, the general preference is not to talk about the yeti. In contrast, among the lay people there is only to a very small extent a taboo against talking about the yeti and observations of them.

According to Stonor: "To see the yeti or hear his cry is considered a very bad omen. People will usually perform a small ceremony to ward off any threatening disaster shortly after an incident. Apart from that, the people did not ascribe the yeti with any special properties compared to known animals... All Sherpas will tell you, that people who see a yeti are absolutely terrified. Their attitude is reminiscent of the Indian peasant's attitude towards the tiger." Interestingly the Indian author Gupta writes something similar: "Every Sherpa believes in the existence of a *de-ti*. Recently about a thousand Sherpa pilgrims in Kathmandu were interviewed, and about a hundred of them stated that they were aware of the creature, and ten people testified that they had seen the creature with their own eyes."

A final word on the Nepal-Sikkim information about the Bigfoot is an excellent recent article by the eminent British primatologist Osman Hill. He considers the main body of evidence to be in favour of the Bigfoot, although he will not yet form a definitive conclusion, but writes that the best description so far is by Wyss-Dunant (1952): An unknown plantigrade bipedal mammal can be found in the Himalayas. It lives in small groups. Osman Hill adds, that future searches should be directed towards lower areas – the thick rhododendron and other forests on the lower parts of the mountains.

Bhutan

Professor Y.N. Roerich has offered very interesting information from this area. "Local Tibetans and Mongolians say that to the north-east, around the area of Dzhade, where the forested valleys start, you can find wild men. They have almost no neck, so you get the impression that their chin sits directly on their chest." "We have also received information from another area at the border between Bhutan and Tibet. Only a limited number of people have visited this area. I talked to several Mongolian pilgrims. Two of them had actually visited Bhutan. They told me that in this border region they had seen a very large monkey. I asked them to describe what they had seen, and their description reminded me of an orangutan. In any case, it was not the Himalayan langur with its very characteristic long hair. As they walked through the forest, they met with several of these ape-like creatures. They said the monkey is so large, it can be mistaken for a person."

Desmond Doig wrote in one of his letters: "In Bhutan, the Western interest in the yeti is absolutely incomprehensible to the natives. I was able to gather a large quantity of folkloric data about the creatures. I am one of the few foreigners who have visited this place. Two people had actually come face to face with the *e-gyo* while looking for a special tree that is used for tableware in Bhutan. The two persons and the yeti fled in opposite directions."

Indo-China

A very famous lama Tsultung Zangbu said that when he was in seclusion in the Chari mountains in Assam, he met a yeti face to face. According to his description, the wild man is an ape-like creature with the body covered in 4 cm long hairs and a swarthy complexion. In 1954 a Belgian anthropologist found humanoid footprints at high altitudes in the mountains of Burma. He considered them to be from a primitive man and not modern man. He also found a number of large stones piled on top of each other, which he considered to be the work of the yeti. According to P.B. Singh, head of the Indian group, who visited the area in 1956, in the foothills of Kubra in Manitsure, near the border between Assam and Burma, two members of his group watched a thin hairy brown creature, about 8 feet tall, with black hair hiding its face. It should be stressed that they, according to the report, actually saw the creature in a Naga village, and that it must therefore have been tame.

In southern Burma, near the Thailand border, the American traveller H. Davis (1940) heard about the *kung-lu*. This is a large creature resembling a gorilla. It lives in the mountains, but only comes down very rarely, allegedly to take people, although nobody could remember any specific cases.

For northern Burma Sanderson reports information given to him by an American soldier regarding the *current* – another name for the kung-lu. This was described as a person with an excessively big mouth. His informant told him that in two cases he had actually grabbed hold of a *current*. Both times, the animal came out of the jungle on a clear moonlit night, apparently in search of food. In both cases the young man thought it was a thief and tried to detain him with his bare hands. The animal did not attack, but broke free easily and fled into the jungle. All it left behind was some long, black, shiny, coarse hairs in the hands of the young man. In the moonlight he had a good look at the animal. It was about 180 cm in length, had broad shoulders, a relatively small head covered with black hair, and straight legs like a man.

From Laos we have the following information from an official report of the French mission: "...not the first time I heard strange stories about wild people living in the mountain forests. They are to be found in the inaccessible part of the Annamskogo mountain ranges... Judging by the local description of these 'wild men', they are covered with red hair. They have become rare. They haven't been seen for a long time. However, sometimes you will find their tracks." This official document later came into the hands of the writer Jean d'Em, and inspired him to interpret the red-haired wild men as relics of the Cro-Magnon people.

We have a little information from Cambodia as well. A big game hunter claims that he had seen a 9 feet tall creature in the jungles of Cambodia accompanied by a female nearly 7 feet tall, and a 3 feet tall young one. They left the same tracks as the Himalayan yeti, and walked on two legs.

When a Soviet engineer was on a business trip in Vietnam in 1955–1956, he learned from one of the employees of the Vietnamese Ministry of Industry, that "in the jungle, in the foothills of the mountain range that separates Vietnam from Cambodia and Laos, you can find great apes (taller than a human), that look like a human, and move on two legs. Local residents who use the mountain trails to travel to Cambodia often meet these monkeys." It was further reported that the only way to escape from a meeting with this creature was to put on a bamboo arm cuff. Rumour has it that when grasping a human's hands the creature begins to laugh, throws his head back and closes his eyes. This goes on for so long, the person usually manages to escape. To verify this information we turned to Professor Chan Hyuy-Leu, the director of the Institute of History at the State Committee of the Democratic Republic of Vietnam (Hanoi). The latter stipulated that, although not an expert in these matters, the following information had come from conversations with the natives of the Tai Nguyen district, the mountain border area adjacent to Cambodia and Laos. "According to the stories of the inhabitants, the forests and mountains in the Thai Nguyen district are home to a human-animal, which people call the *an-nak-tan* or *zo-uate*. According to the comrades currently working at the radio station Voice of Vietnam, a young man killed an an-nak-tan with a crossbow and a poisoned arrow in 1944 when it was coming down to the creek for food: crabs, shells and seaweed. The wounded an-nak-tan ran screaming into a mountain cave on two feet. When the body was found, it was a pregnant female the size of an 11 or 12 year old girl." Professor Chan says he has also talked with a few educated persons with positions of responsibility in Hanoi (Comrade Ma-Khe, who is currently the director of the School of Employees of National Minorities in Hanoi, with fellow Wu-Wang and I-Ngong , deputies of the National Assembly of Vietnam), who confirmed that local people often see a hominid animal they call zo-uate or an-nak-tan.

All information relating to the Indochinese Peninsula is still very sketchy. Furthermore, we do not have any reliable information regarding populations of such creatures further towards the south, in the Malay Peninsula and the islands of Indonesia. There are a few reports from Malaya. However, from the southern part of the island of Sumatra, there is a lot of information. According to the local population and travellers the area is home to a creature known as *sedapa* or *orang-pendek*. The presentation of this material would amount to a whole new chapter.

CHAPTER 8 • INFORMATION FROM THE MOUNTAINOUS REGIONS OF SOVIET ASIA

Pamir Mountains (East Tajikistan)
In the last decade the Soviet Pamirs have been very quickly populated and economically developed, leading to major shifts in the local fauna, especially large mammals. The passage of large numbers of people through the Pamirs since the 1960s has greatly changed the local nature. The large herds of wild yaks have completely disappeared, and they are supposedly the main food of the wild hairy men – *yavan* or *aban*. The most profound changes in the fauna of the Pamirs have happened in the last decade. So when speaking of the Pamirs we have to consider whether we can assume these creatures are now extinct, even though it may have happened quite recently.

Some information regarding the wild man in the Pamirs reached Russian educated circles before the First World War. We have also found information from G.K. Sinyavsky: "Before the Revolution I heard from officers of the Pamir detachment, that the wild men sometimes have tried to attack them." Sceptics however explain the tracks found as those of the snow leopard. There are even rumours, although not verifiable, that in 1913 Russian military in the Pamir killed a wild man, and that a doctor dissected the body. The Russian Geographic Society and the Russian Academy of Sciences then promised to send a special expedition to Pamir, but it was stopped by the First World War. In 1914 a small expedition did leave for Pamir, but they had ethnography and linguistics as their main subjects, and we have not been able to establish whether they had any connection with the wild man problem.

The first reliable data are from the time after the establishment of Soviet power in the first half of the 1920s. In 1924, Doctor S.I. Sour came face to face with a humanoid female covered in red hair on an abandoned mountain trail of the slopes of the Darvoz Range near the Obikhingou River. Both were frightened of each other and they fled in different directions. Later, Sour hypothesised that the woman he met was a descendant of the lepers, who are expelled into the mountains.

By the autumn of 1925 M.S. Topilsky, who is now a retired general living in Moscow (born 1901), was leader of a cavalry regiment battling the Basmachi. [45] He has supplied us with the following story:

> We were pursuing the gang, which was operating in the Western Pamirs. From Khovaling we went on to the district of Darwaz Garmo, turned south and crossed the top of the Vanj and Yazgulem Ranges and went on to Rushan. In the highland villages of Vanj we have heard stories of hairy man-beasts, monstrous creatures that live in the mountains, but not only in the snow. They do not usually attack people, but when confronted on a mountain

trail they can kill, and will rip off a person's head. According to legend, meeting with the creature will bring misfortune and death. The high mountains are their kingdom, and humans cannot go there. Nothing lives there, not even rams or leopards. Only one of the people we talked to claimed to have seen such a creature. But according to the hunters, they often hear them howling and screaming, and the residents of the villages can distinguish the voice of the *poluchelovka* from a human or another animal. We were shown a Muslim holy cave high in the mountains with a shrivelled corpse of a sitting holy hermit. One could only perform the pilgrimage to the shrine at certain times of the year in the summer. The rest of the time, the mountains are ruled by the man-beasts. We did not attach any particular importance to these stories.

Before we started pursuing the gangs in the Pamir Mountains, we were warned that we were entering a territory with man-beasts. One day on a mountain trail, when we were tracking a gang, we saw footprints crossing the trail. We had good local dogs with us. They sensed something, but would not follow it. The tracks were very clear, and looked like bare human feet. The track was more than 150 metres long, and ended at the foot of a steep bare rock. Our medic examined the tracks carefully and judged them to be human. He also found evidence in the form of feces. It looked human and was very dry, consisting of the remains of dried berries.

We continued and overtook the remnants of the gang in a place where there was a glacier and a very narrow trail. In the rock beneath the glacier there was a small cave. We surrounded the area, and on a hill above we installed a gun. When I threw a grenade down, a man ran out onto the glacier and in Russian (he was a Russian officer, who was part of the gang), shouted to us that the ice would collapse and bury everything. He asked for time to warn his friends and fled back to the cave. Soon we heard the sound of ice moving. From the bottom also came the sound of what we assumed was shooting. Fragments of ice and snow started falling and covering the entrance to the cave. Three people managed to get out before the entrance was completely blocked. The rest (five persons) were buried in the collapse. Two of the three people were killed by our fire. One was seriously injured. When we went to him, he showed us where the collapse had been. He also told us that his gang had met up with hairy humanoids in the area, when they went into the cave the first time a few days before. There were several of them, and they were all screaming. They had sticks in their hands and attacked the gang. They tried to defend themselves, but one of them was killed by the man-beasts. He himself had rushed to the exit, pursued by one of the monsters, but it was shot when it came out, and was later covered with snow.

We dug around in the snow, and found the body with three bullet wounds.

Close by I found a very strong stick of wood, but I cannot be certain it belonged to the creature. At first glance it seemed to me that I was looking at a monkey, as it was covered in hair. But I knew that there are no apes in the Pamirs, and the corpse did look like a man. We tried to pull at the hairs of the fur, to make certain it was not a man using a skin for camouflage, but it was natural hair. We turned the corpse around and measured it. After careful examination, we concluded it was not a person.

The corpse was a male, height 165–170 cm, possibly an elderly one or even old, judging from the colour of the hairs in some places. In general the hairs were grey-brown, but on the chest they were brown and lighter on the stomach. In different parts of the body they had different lengths and thickness. On the chest they were longer and denser, on the stomach they were shorter but thicker. In general the fur was very thick, though without an undercoat. The least amount of hair was on the buttocks, so the creature probably sits like a man. The most hair was on the hips. There was no hair on the knees, but callouses from kneeling. Shins were less hairy and almost bare at the bottom. Feet were naked and covered with rough brown skin. Shoulders and arms were covered with hair, the density of which is reduced at the wrist, and on the back of the hands. The hands were completely without hair. The skin rough and calloused. The neck was hairy, but the face had no hairs. Complexion dark. No beard or moustache, and just a few hairs on the edge of the upper lip. On the front of the head the hair was thin, but at the back it was very thick and matted. The dead lay with eyes open, and teeth bared. Eye colour is dark. Teeth were very large, flat, but shaped like human's. Sloping forehead. Above the eyes were very powerful eyebrows. Strongly protruding cheekbones, giving the face a similarity with the Mongolian type. The nose is flattened and wide. The ears were hairless, and seemed somewhat more pointed at the top and with a longer lobe. The lower jaw was very massive. The dead one had a powerful and heavily muscled broad chest. We could not see any difference from man. The sexual organs were like those of a man. The hands were of normal length. The fingers and toes were also studied.

We do not know exactly where this took place, and we had no precise maps of the Pamir highlands. We can assume it was between the Yazgulem and Rushan ranges. The nature of the dead creature is essentially a mystery to us. But it was impossible to bring the corpse in. It was heavy, and the path was difficult. We could also have difficulties with the locals. We discussed removing the skin, but it would be like skinning a person. In the end we decided to bury the corpse at the scene.

We went further south, and at the first opportunity went down the mountain. The Baluchi inhabitants of the area took us for some sort of expedition. They were surprised to see us coming down from an area considered to be the home of humanoid monsters (I do not remember the local name) and therefore

> inaccessible to people. We got a lot of new information from the Baluchi people. We were told that the humanoid hairy creatures are not only to be found one at a time, but sometimes in pairs or three of them including a baby. They do not occur in large groups. We were introduced to one of the Baluchi, who is considered to be crazy as a result of a close encounter with the creature. He told us that one day, while hunting with a friend in the mountains, they came upon a wild hairy man. They tried to shoot him, but it threw a large rock at them, and killed the man's friend. The hairy man then fled, and left the other, who lost his mental balance. In one of the villages they told us that in the recent past hunters found a cave with lots of argali bones. The clan elders explained that this was the camp of the monsters, and they imposed a strict taboo on the cave and forbade the hunters to visit these places.

There have been two expedition to the Pamirs, a geographical one in 1928 [46], and the Tajik-Pamir expedition in the 1930s. However, we only have scattered information from them. The geologist and mountaineer L.L. Barkhash was a member of a group headed by N.V. Krylenko. In late August 1929 he saw fresh tracks on the Sauksara Glacier at an altitude of 4400 m. They were found in a small shaded place, where snow had fallen during the night, and not yet melted. They were like fresh tracks of human feet, going up the glacier. The group decided Barkhash had seen bear tracks, but he did not agree with that. To the west of the glacier were some gentle, almost horizontal moraine landscapes covered with grass. The area had plenty of marmots and sometimes herds of wild goats. It is possible the animal was going there to hunt. On another occasion in early September 1933, in a camp on the Turamos glacier, at an altitude of 4200 m, there was a lot of noise and crying from the animals, and later he found fresh tracks. In the morning he heard a strange sound, like a howl, and not the desperate cry of a man. The sound was repeated twice, and it was very different from all other animal sounds he had heard in the Pamirs. One of their porters, Tajik Yusup, claimed it was the sound of a wild man. In September 1933, at a height of 4300 m they made an unexplained discovery one evening. In the fresh grass they found the body of a marmot without a head. It was separated from the vertebra as if cut with a razor. There was not a drop of blood on the grass, and no tracks. What animal could bite off the head of the marmot, drink all the blood and discard the carcass of the marmot like that?

When the academician D.I. Shcherbakov in 1933 was crossing a pass from the upper Vanj region, he saw footprints in the snow resembling a human. They were clearly different from bear tracks. In 1934 the climber V.N. Markelov, a member of the group led by N.V. Krylenko, was near Lenin Peak at an altitude of about 6000 metres with his friends, when they noticed two figures at the bottom of a ski slope move in a bipedal upright position at a distance of about one kilometre from the climbers. They quickly tried to descend to the unknown figures, but they soon disappeared from sight, and the climbers could not find them. According to Markelov there was not supposed to be any people on Lenin Peak at the time.

In 1936 an observation was made by G.N. Tebenihin, an employee at the radio-meteorological observatory at the Fedchenko Glacier. According to him, an event in March 1936 was never fully explained. One day in early March B.G. Nelle was on duty at the meteorological site. During his morning rounds he noticed that one of several two-metre strips mounted in the glacier to monitor its movements was out of place, and that the snow around it had been compromised. The team decided it must have been done by a bear, but Nelle and Tebenihin went skiing on the glacier to look for themselves. They brought guns as well. They found there were also several rows of rods, which had all been broken and left on the ground. "We could see the track where someone had moved across the glacier to the rods. We carefully examined the broken bits, but could find no trace of teeth or claws, and there was not a single hair in the snow around them. Only lots of footprints, and the tracks were very fresh. However, due to the consistency of the snow, we could not see any details, but they did not look like bear tracks." They were also longer and wider than his size 43 walking boot [UK size 10 ½–11]. A photograph was taken with skis and a rifle as measurement. Only one image was attached to the report received by the committee. There was a creature moving away in the distance. Both men tried to follow the creature, hoping for it to be a bear that could give meat for the winter. The tracks showed clearly that the creature had walked on its hind legs all the way. They tracked the animal for several kilometres up and around the glacier. At one place the creature had sat down, and they could see the prints of its feet and buttocks. Later they found its tracks on the Nalivkina Glacier. In the end they had to abandon their pursuit. They had followed the creature for at least 15 km.

Tebenihin thinks that this incident explains some of his other observations and events that took place during the winter of 1938–1939 at Altyn Mazar, a few kilometres from the Fedchenko Glacier. This refers to, for example, the fact that when people are producing firewood for winter the horses are always frightened at certain locations.

At a later date we have found Nelle (now head of the assembly of a construction site in the city of Tashkent) and have interviewed him about this incident. His explanation is very different from Tebenihin. Nelle worked at the observatory at the Fedchenko Glacier in 1936 as a senior weather observer. As he described the events, they took place in the beginning of May 1936, when the snow was melting. The stories differ in several minor aspects, but the most significant is, that it was all in the morning, there were no clear footprints, and the creature was visible at a distance of about 1 km for the time. Nelle thought it impossible for the pursuers to determine whether they were tracking a "bear" walking on its hind legs or on all fours, as it was sinking into the snow, and they were on skis. Several times the creature stopped and sat in the snow. This left a clear mark of the animal's bottom. When the creature was sitting down, Nelle fired at it with his rifle but missed because of the distance. Visibility was good the whole time. The "bear" was dark brown. If it was a bear, it was the biggest Nelle had ever seen. Judging from the tracks they found, its legs were unusually large for a bear. The beast was watching the pursuers all the time, and only when they stopped would it sit down to rest. The "bear" had come to Altyn Mazar from the glacier. It could have spent the night on the southern slope, where the snow was already gone. It is an area rich in wildlife, with goats, leopards and rabbits.

According to Nelle, a geophysicist called A. Kozhevnikov was watching almost the entire pursuit through binoculars from the observatory. If we could find him, perhaps he could resolve the discrepancies. Nelle is inclined to believe they were chasing a bear, but that does not fit in with the many incidents of the beast sitting down on its rump.

In 1938 geologist A. Shalimov and a group of Tajik porters were passing through the Vanj Range. After spending the night at the foot of the steep rocky slope that leads to the Gudzhiva Pass, they made a halt at noon on a snowy saddle. The porters suddenly discovered that they were surrounded by fresh tracks, made this morning by “the wild man”. The whole chain of porters rushed down the snow-covered slope with extraordinary speed. It is important to emphasise, that they have no problem recognizing fresh bear tracks. The geologist stayed behind to examine the mysterious footprints. “There were distinct traces of bare feet on the blue surface of the sparkling fresh snow. The footprints led from the valley on the northern slope of the ridge, and disappeared on the rocks, where the wind had blown the snow away. I made a drawing of the tracks in a field diary. The track length was about 30 cm, the width 15. There were clearly visible imprints of five toes. The first toe was much larger than the others and was sticking out to one side.”

Geologist B.M. Zdorikov, who worked in the Pamirs from 1926–1938, has communicated some very interesting information. Zdorikov, who was leading the work in 1929 in the western Sanglakh, was also asking the residents about the local fauna. According to Rais, the chairman of the village council of Tutkaul village gave him the following list: wild boar, bear, red wolf, porcupine, jackal, hyena, and *deva*. The geologist was surprised that Rais did not consider deva to be evil spirits, but animals like the wolf and wild boar. However, he said very specifically that the deva was built like a small but very stocky man, walks on two legs, and has a head and body covered with brown hair. They are very rare in Sanglakh, but they can also be seen in pairs, one male and one female. Rais had not seen young. Last summer an adult was caught alive in a mill, where it was eating flour and grain. It was just a few kilometres from Tutkaul. The deva was held captive for two months on a chain at the mill. It was fed with raw meat and bread. One day he broke the chain and fled. Zdorikov was also shown a man who had a large scar on his head. This was allegedly from a deva.

In 1934 Zdorikov himself came upon the animal. Together with a guide he was making his way along the Tajik alpine trails through the thicket of wild *grechika* [usually meaning buckwheat, but in this context presumably some sort of rough shrub] on the mountain plateau at an altitude between 2500 and 2800 m above sea level, in the area between the Darvoz Range and Peter the Great Range. Suddenly, he says, they saw a small area where the grass was trampled and the earth dug up. There were visible bloodstains on the trail and scraps of marmot hair. "And right at my feet, on a mound of freshly dug earth, a strange creature was lying asleep on its belly. It was stretched out to its full height of about a metre and a half. The head and front legs I could not see well. They were behind a wilted *grechika* bush. I caught a glimpse of legs with black bare feet too long and slender for a bear. The creature's back was also too flat to be a bear. All of the animal's body was covered with shaggy hair, more like yak wool than fluffy bear fur. The coat was brownish-red, more red than I had ever seen on a bear. The chest of the beast was rising and falling slowly and rhythmically. I froze in surprise and looked around in confusion on the Tajik following me. His face was pale as a sheet, and he pulled at my sleeve and silently beckoned me to flee immediately. I have never seen such a look of horror on a person's face. The fear of my companion was transferred to me and we both ran without looking back, stumbling and falling in the tall grass. The locals were very concerned and alarmed by the encounter with the sleeping deva. According to the inhabitants in the valley, there were a few families of devas living in the mountains. They were considered to be beasts, and not seen as representatives of evil forces. They usually do not attack humans or livestock, but it was still considered bad luck to meet them."

Another geologist, S.L. Klunnikov, was important in the early studies of this unknown species. Klunnikov wandered through almost the entire Pamirs on foot, and collected and compared stories about the great unknown apes on the way. He heard so many stories that he had no doubt there was a mysterious animal living in the Asian highlands. Several years ago, a hunter family even showed him a tuft of coarse grey hairs cut off from the body of one creature killed by his grandfather. The hunter would not give up the hair, as he considered it a talisman which should be given from father to son in the family. Unfortunately the many data collected by Klunnikov have been lost, as he died in the war.

In August 1958 another geologist and expert in the Pamirs, S.I. Proskuko, made an interesting observation, and photographed fresh tracks like the ones found by Eric Shipton, at the South Alichur Range, at an altitude of about 4500 m. The heel had made a very clear imprint, the length of the foot was 18–20 cm, the second and third toe seemed to be fused, but the big toe was clearly defined. The distance between two tracks was 65 cm. The band of snow was very narrow, so there were only two right and two left tracks. They went down from the glacier. But higher up, near the glacier itself, Proskuko saw a chain of similar tracks aimed towards the glacier. "Judging by the prints," he wrote, "the animal was walking hard on two legs...And although the snow had melted, the prints of the big toe were very distinct. The tracks looked like the ones I had seen below. They belonged to a biped. The tracks went into the valley in an area with many stones, and here I could not follow."

In our published "Information Materials" we also have statements of former administrators and former employees of the Border Guards. For example, an employee of the Al Maluya administration, who worked in Vanj for six years, heard a lot of stories from the people of the villages in the Yazgulem Valley, especially the older generation, about the existence of some form of humanoids at high altitudes, particularly in the upper reaches of the Yazgulem near the Fedchenko Glacier and in the direction of the Bartang area. Such stories usually came from hunters looking for wild rams and snow leopards. Such stories often mentioned household items disappearing from tents, and the hunters then finding them in the mountains. It is interesting that the administrator never heard about the existence of Bigfoot in other areas of the Pamirs.

A report from former border guard P. Pyankov comes from a completely different area of the Pamirs. "In mid-summer 1932 I was part of a squad of seven people in a pass leading to Kashgar. In the afternoon, we saw a man who was moving on the snowy mountain pass towards the trail. We prepared to arrest him, as this was a violation of the border. The 'man' was well lit by the setting sun. We saw him clearly at a distance of 600–700 m. He moved with a quick easy gait. He was of medium height, stocky and bent forward with long arms. He crossed the path at a right angle, and headed into the impenetrable rocky area. We went to the place where he crossed the path. On the snow we found tracks, both human-like and non-human. They were not shoe- or boot-prints. The 'man' was walking barefoot. The tracks led into the rocks, we searched as much as we could, but found nothing. Then we went along the tracks in the opposite direction and found that the 'man' had walked along the ridge, which forms the border and should be out of the reach of humans."

In the summer of 1960 one observer, Alex Grez, reported finding tracks in April near Lake Zorkul, in the Kara Djilga canyon (altitude about 4200 m). To the west the site is adjacent to the eastern end of the Wakhan Range, and to the south lies the Wuhan-will Valley behind which lie the glaciers of the Hindu Kush. On the western shore of the Chakan Kul lake, Alex saw a strange set of tracks he had never seen before. The tracks were in a thick layer of snow. They went vertically down the very steep western slope of the gorge, and crossed the lake at its narrowest place, turned south and went up the gorge. Alex tracked it for more than half a kilometre, but then had to go back, because he had no goggles and was beginning to go snow-blind. Alex later made sketches of the tracks. He states categorically that the tracks have nothing to do with bear-tracks, which he knows well. The tracks were broad and narrowed towards the heel, and there were clear and distinct folds on the underside. On one there were clear toe-prints and claws, but not sharp like on a bear, more rounded like nails. Each print was 25–27 cm in length. The creature would move on two legs, except on slopes, where it would move on all fours. Unfortunately Alex did not sketch the forelimbs, although he said they differed from the back legs. The step length was longer than the step length of Alex Grez, who is slightly below average height. In areas of loose snow it was the same. The legs of the beast were set wider apart than in a man. In several places, the animal had stumbled on one foot, and left knee-prints. After trying to fall in the same way, Alex Grez noticed that the shin of the beast must be very long.

We also have information from the late hydrologist A.G. Pronin: “In 1957 I stayed for a few days in the Balyandkiik Valley near the Fedchenko Glacier. On the 12 August I was examining the river valley before noon, when I saw a curious figure. On the left side of the valley (about two kilometres from the mouth) I saw a moving figure resembling a human. It was about 500 m away. The figure was stopped, the legs set widely apart, and the arms were longer than an ordinary man. Visibility was excellent, but it was not possible to see the fur. After five minutes the strange figure disappeared behind a rock. A few days later I was passing the same place just before sunset, when I saw the same figure again, but only very shortly, as it disappeared into something black, perhaps a cave. I believe the animal saw me.” Pronin also reports on another incident. In early September 1957, an inflatable rubber boat went missing from the mouth of the river. It was later found five kilometres upstream, although travel by boat up the rough mountain river was impossible.

After Pronin's story was published in the newspapers, it attracted a number of responses, with information about similar observations. Some are very brief. The artist M.M. Bespal'ko reports that “I saw this same being in the Alichur Valley on the 29 July 1943” where she was doing sketches. A former border guard wrote: “After the war, while serving in the area, we were three soldiers and a sergeant, who saw a group of strange two-legged animals. There were three. They looked similar to what you wrote in the newspaper. Because of the great distance, we could not make out any details. We found no tracks, and the dog would not follow the trails." A group of former guards wrote: “In 1954 we served in the area of the Fedchenko Glacier. We have seen the 'Bigfoot'. We believe it is necessary to look for it in the summer. It is difficult to catch, it is cunning and cautious.” According to verbal information from the mathematician A.D. Alexandrov, who is a member of the Soviet Academy of Sciences, he found footprints similar to the pictures in the newspaper in August 1961 at the clay bank of a mountain stream near the village of Vrang Nishgar (Nicholas Range).

In the summer of 1961 in the Regarskogo district Boboev Tour confidently told us that to his knowledge a year or two earlier, that is in 1959 or 1960, the military caught two *odes-yavoi* (wild men), a male and a female, and took them to Dushanbe. Their bodies were covered with hair, they did not speak, and would eat only raw meat.

In October 1956 at a meeting of the Eastern Committee of the Geographical Society of the USSR, I. Sakharvo made a statement saying that, based on the stories of the Kyrgyz in the Eastern Pamirs, we can assume that humanoids are living in the remote areas of Pamir. They live in small families, are very few in number, and avoid contact with humans. Outwardly they look like some mix between man and ape, overgrown with thick hair.

In 1957 K.V. Stanyukovich published an article with information about the Bigfoot in the Pamirs. Subsequently in 1958 as the leader of the Pamir Expedition he wrote that "for the last two years I've asked old Pamir inhabitants about the wild man based on these earlier inquiries. In the most remote and totally uninhabited Pamirs, in the valley of west Pshart, lower Murghab and a number of rivers flowing into Lake Sarez in the south, as well as in the area of the lower Balyandkiik, Kaindy and Sauksay (near the Fedchenko Glacier), a number of shepherds and hunters have met a wild man. The wild man is about the same height as an ordinary man, but is completely covered with hair, except for the face. He uses no fire and does not know what a gun is, but he can throw stones and sticks. He avoids people and feed on roots and small animals, which he catches or kills with stones (hares, marmots). In winter he can kill argali sheep in deep snow. He moves quickly and probably does not have a permanent home. He is very rare, and was met with more often in earlier times." After the 1958 expedition he changed his mind about the wild man, and decided that the people were retelling an ancient myth. These stories had misled the people to think that the "Bigfoot" still existed. The researcher D.N. Evgenov has also gathered information in the Pamir about the *biyavan* and, in particular, stressed that according to the population it was more common in the 1920s than it is today.

In 1958 experienced Western Pamir hunter and pathfinder Mirzo Kurbanov told about an incident in the sheep pen in the geological camp near the glacier. Due to snow leopard attack on a flock of sheep, there were hunters on the lookout. Late at night, when they were sitting around the fire, they heard a strange scream coming from far away in the upper reaches of the valley near the ice. It made the shepherds extremely anxious. They said it was the wild man who lives in the mountains. According to the shepherds, the wild man looks human, but is smaller, has no clothes, is covered with hair, very strong, runs fast. Interestingly, the shepherds ward off the wild man the same way as the monks in the Nepalese monastery in Tengboche, by making a lot of noise.

A resident of the village of Kantadash said that in 25 years of living in Central Asia, she had heard a lot about wild people living in inaccessible places in the Pamir Mountains. Sometimes hunters would meet them, or they would show themselves to people who were lost, but then quickly hide. Sometimes they would take things from sleeping hunters, but always leave another thing in its place.

We also have an interview with a Kyrgyz hunter named Kerimbekov about his observations of the "big man". The last time he met him on the South Alichur Range above Lake Zorkul. He has no clothes, and is covered with hair, except for the face. The arms are very long. His father had told him, "that this man runs like an argali, can see like a golden eagle, and hear like a leopard." The first time Kerimbekov saw the creatures there were two. It was daybreak, and they were moving quietly down the hill in the valley, where he had spent the night with his father after the hunt. However, the movement of the horse scared the creatures and they climbed back up. The second time Kerimbekov saw one of the creatures he was waiting on one side of the gorge for the female mountain goats. When the goats came in the evening, one of them was hit by a hail of stones and fell down. And then a hairy man went down after it, and dragged it into the high bushes. We must be cautious with this story, as it does not have any close parallels in our archives.

The stories recorded by K.V. Stanyukovich are much simpler, more realistic and resemble our huge set of data. In 1936 local workers refused to spend the night at the Sauksay River crossing, saying that a wild man was living in the area. A Kyrgyz reported at the same time that some time ago she had seen a wild man near the mouth of the Sauksay. She hid in the rocks, and suddenly the wild man started running up the hill and screaming. In 1937 there had been a lot of talk among the Kyrgyz around the Togar-Kata Pass, that a *gul-biyavan* was living there. A Kyrgyz named Dzhemagul said that a long time ago he had seen from a distance two "wild men". They went up the hill, where they "dug the earth and ate the grass" – probably roots of some kind. Dzhemagul said that gul-biyavan was afraid of people, and would hide from them, but even if he did attack it was nothing to be afraid of. You just have to shout, and then the gul-biyavan would leave. In 1951 people on a collective farm were warned that three wild men were living down river. They said the wild men were covered with hair, could walk very fast in the mountains, usually kept themselves hidden, but could sometimes get angry and attack the lone traveller, or challenge him to a fight with shouting and punching the ground.

A small reconnaissance expedition from Leningrad University in the spring of 1958 collected a variety of data among the Kyrgyz in the Eastern Pamir. In particular they were told by a 60 year old hunter named Tusupov that the track of the gul-biyavan was like human tracks. There were no visible claws, the big toe would bulge out slightly, and judging from tracks in sand the foot was hairy. A young Chattukoy hunter found tracks of a gul-biyavan in fresh snow, and followed it for a long way, but never saw it. In the same area two other Chattukoy hunters found tracks in wet ground. They were shown pictures of Bigfoot and found them to be very similar.

O.E. Agahanyants, a researcher at the Pamir Botanical Garden, recorded a number of stories about wild hairy men in the Western Pamirs, including stories from the Wakhi people of the *almasty* - a woman, covered with hair, with breasts so long that they dragged along the ground. Shepherds would also guard the cattle at night from the wild people (like almasty, but men) in the Bizhon Gorge on the western slope of the Shakhdara Range. Near Irkht (at Lake Sarez) there were stories of wild men living in the mountains. They were hairy and strong, and roamed freely where they liked. In the middle reaches of the Bartang one person gave the same information. He said: "Before the Revolution there were wild men, and now there are not." In general, said Agahanyants, it seems that most of the informants believe in what they say, though sometimes it is superstition.

The next part is information collected during the Pamir Expedition in 1958. They are very varied. Some are concise but detailed, others are of a more semi-legendary character.

Let us start with those who claim to have seen a wild man with their own eyes. Kadir Tokoev, 58 years, Kyrgyzstan, working at the Bulun-kul farm, said that he saw a gul-biyavan in 1922 or 1923, when travelling with five neighbours in Alichur Tokhtamysh. The road they had chosen was desolate and difficult. On the way to the Belez Pass, the travellers were warned of the gul-biyavan living there, but they were not afraid. Around noon, all six men saw something that looked like a human being without clothes, but taller and with grey fur, coming down from the top of the mountain. Kadir Tokoev went forward, intending to shoot, but when the gul-biyavan was still 500 metres away, he disappeared into the bush.

Abdilda Dzheembekov, 58, Kyrgyzstan, who was working on an archaeological expedition, says that when he was 15 years old he went out with the hunters and their dogs in Mardjanay. "One of the dogs went off and started driving some animal that looked like a man. Kyrgyz hunting dogs will not drive a normal person, but only a wolf, a bear, a dog or another animal. The running beast was about 400 m away. It was the size of a normal person, it was grey. The hunters were convinced it was not a wolf or a bear." Yusup Palvan Sadykov, 74, Kyrgyzstan, farmer-shepherd, in 1945 saw several unfamiliar five-toed footprints in the sand in the area around Chattukoy. The prints led to the mountain. Foot length about 35 cm, width about 20 cm. There were visible imprints of nails. The distance between tracks about twice a normal human step. Aaryn Abduraimov, 67, Kyrgyzstan, working in Murghab, three years ago in Chattukoy saw human-like tracks in moist clay, length about 35 cm, lots of tracks, descending from mountain top and leading to the water. Turganbaev Shaimkulov, 44, Kyrgyzstan, former chairman of the collective farm, in 1948 found a set of tracks crossing the trail in front of him near Lake Sarez. He was passing the place for the second time, so knew the tracks were less than one hour old. Shaimkulov is an experienced hunter, and was certain the tracks were not bear or dog, but had the shape of human feet. But no man would walk barefoot in February. Track length about 35 cm, width 12–15 cm, first toe bigger than in humans, claw marks could be seen. Hunters started following the tracks, but after a while it started to zigzag. It scared them, and they stopped following.

Some of the records are very short, but still interesting: "The old people told me, that in the old days people would meet the gul-biyavan in the gorge near Sulystyg Tokhtamysh." Sometimes the messages are very concrete and refer to a number of still-living witnesses: 40–50 years ago, a man named Kulata was hunting in the Sulystyg Valley, when his horse suddenly took fright. Further along he saw a gul-biyavan (or gesu tyrmak – which is the same). It was built like a person, but taller, covered with hair, and walked with a stoop. Kulata wanted to shoot the creature, but his frightened horse galloped away. Another witness said that one day, a tall "man" came to a yurt. He was naked and his whole body covered with hair. The hair on the chest and the knees was shorter. Both the humans and the wild man were afraid of each other, so the wild man went up into the mountains and shouted something on the way. A very respectable elderly Kyrgyz near Murghab told us that according to witnesses, in the old days ("under Nicholas"), during one year there were "wild men" in the area between Mammazairom and Kara-Suu. They were called gul-biyavan or gesu tyrmak. They looked like humans, but were covered with short hair. They walked on two legs like humans, but had feet twice the size. Their voices sounded like humans, but were very powerful, so they would carry over the mountains. The people in Murghab would hear the cries of the wild men and later find the footprints. One person, the late Rismet Isakov, drove past a place in spring where people had been digging a pit for construction of a house, but then abandoned it. It turned out that a wild man had used the pit as his den. When Isakov passed by, the wild man climbed out of the pit, screamed and ran up into the mountain. All summer people heard his scream coming from the mountains, but in autumn all was silent, no screams, no tracks. Nobody knows if the gul-biyavan had died or gone somewhere. According to one story a shepherd named Turdybek was attacked by a gul-biyavan near Ak-Beit: "He was covered with short hair, like a camel's hair." They fought hand to hand, then the gul-biyavan disappeared, and the frightened Turdybek returned home and led others to the scene and showed them the marks on the wet and soft ground; the tracks looked like bare footprints.

Situations similar to those described occur again and again. A big man, covered in hair, came close to the tent, and even stuck his head into it; he was overgrown with hair. Then he jumped back and ran away. This was in the Dzhandaban Gorge. Again and again we hear from people who have heard about the existence of the gul-biyavan two or three decades ago. A hunter called Otunchu Maatova said that twenty years ago he saw a gul-biyavan in the Urus Dzhilga Gorge. The hunters had binoculars, so the "man" was easy to see. He was walking down the slope. He was covered with small greyish hairs. He had a great (shaggy) head, and wide upper body, and he walked a little stooped. One evening in the Kurteke area, a Kyrgyz named Tursun-amine saw a humanoid creature coming down the mountain making strange noises. A Kyrgyz named Toktosyn Sarykulov was in a narrow rocky gorge in ancient times ("in the time of Nicholas") in the autumn, when he saw a large creature that wasn't shaped like an animal, and a man that had no clothes and was covered with greyish hair, resembling the colour of a new camel coat. The creature stopped and stared at him. Gul-biyavan was last seen on the road between Ar Chirac and Kara-Kiya in the Aksu valley. It is said that a Kyrgyz named Chumukbay who was searching for a missing camel saw something similar, but when coming closer found a sleeping gul-biyavan. "He was lying face down, breathing like a sleeping man. Suddenly he woke up and disappeared." In 1939 Turdybek Baysaryev, according to his son in law, was attacked and seized by a gul-biyavan. It was covered with short hair all over. It had very short hair on the face. It gave off a very strong odour. In Western Pamir a hunter saw a female wild man between the Roharvskim Gorge and the Bodauli Gorge. It was a little taller than a man, with hairy arms and legs, shaggy head, a little chest hair, long breasts and bare face.

We have now been through quite diverse descriptive material relating to the Pamirs.

Most of the observations date from the past. There is very little recent material. Much of the information is second hand. And many of the reports talk about the superstitious fear of illness because of a meeting with a gul-biyavan or from just talking about it. It is our impression that this superstitious fear is supported not only by local traditions, but also from active influence of the Muslim clergy. This writer spoke with Mullah Kurumshu in Saryk Mogol (Alay Valley), an elderly man with extensive ties to the Muslim world and a man of great experience. On the question of gul-biyavan he replied that it certainly exists, but hastened to put it on a par with djinns. He explained, among other things, that the gul-biyavan and almasty are the same creature, but male and female. When I asked him if anybody had seen a live gul -biyavan, he said no. But added: "Today gul-biyavan can be found in some places in China or India."

It is clear that the observations are scattered all over the Pamirs. There seems to be no fixed time for the observations, but there is a concentration of sightings in two areas: the South Alichur Range, and the Yazgulem headwaters and surrounding areas. It is important to note that we had almost no descriptions of encounters with young Bigfoots. The most important are from B.M. Zdorikov who tells us that in the Darvoz Range area there are sightings of whole families of hairy wild men: males, females and young.

Mountainous West Tajikistan, Uzbekistan, Turkmenistan
The Tajik Malik Saidov is a gardener working in Dushanbe. In 1934–1937 he was working for the police in the Tavildara area. In 1935, a farmer from Dashtskogo-Saghir ("marmot valley") had repeatedly filed claims with the village council to stop someone from kicking his tent over at night. In 1934–1936 villagers repeatedly appealed for weapons, as they could not pass the path from the village to the plant (about 1.5 km) at night without encountering the *hum* (wild man). It kept approaching travellers but did not attack. Malik Saidov went to check the path, but saw nobody; perhaps, he adds, "because the *hum* saw my weapons". According to Malik Saidov, it is possible to get plenty of information about the *hum* from the Childara villagers. He also claims that his father-in-law twice in his life had to fight a strong and smelly creature. In 1954 he was found murdered in a strange way: his liver had been torn out from the back. A strange case, especially as the informant is Malik Saidov himself. In 1937 he was appointed head of the summer camp in the territory of Pagolo village in the Tavildara region. The summer camp with facilities for children was built of sticks in a beautiful juniper growth in the valley 3 km from the village, at the confluence of three streams, full of large trout, which are easy to catch by hand. Once on a moonlit night the children shouted that someone was throwing small stones at them. It lasted until the morning. The next night was the same. According to Malik Saidov, he personally saw one boy receive a graze on the temple by a stone. The small stones apparently came through gaps in the brushwood walls, but nobody had noticed them outside. The next night a soldier who had been keeping watch outside with the camp commander was hit in the head by a larger stone. The next night the camp was

guarded by the police, but stones kept falling on the children in the room, and nobody could find the throwers. Local people explained that the stones were thrown by *guls* living in the valley, and they would not allow people in amongst the juniper. They had to move the camp. [47]

According to the Tajik Hidzhobataha Nodirova many have met the *guli-biёbona* (wild man) in the desert and wooded areas around Vahio. Six or seven years ago in the pass between the Karateghin and Vahio, one friend with two companions met a naked humanoid creature covered with short dark hair; a creature that was slightly above the height of a human and gave off a strong unpleasant smell. There are a number of reports about meetings and observations of "wild men" in the area around the Darvoz Range. According to V.A. Hodunova, "surprisingly, the greatest number of stories about meetings with the wild people refer to the region of the Kulyab Mountains, where such meetings are apparently not a rarity. At the same time, there is also a belief that people who meet the almasty (wild man) will either die straight away or a short time after."

Abdulhamid Abdurakhmanov, 67, recorded the story of his grandfather, a hunter named Allaёra. It was in Jirgatol - east of the Karateghin Range. Five hunters went up into the mountains for wild sheep and spent the night in the woods. In the middle of the night, the hunters woke up because their dogs were barking. They put dry grass on the fire and saw at close range a *gul-bo* or gul-biyavan. It was a huge man-like creature almost two metres in height. It was all covered with hair. It gave off a very strong and unpleasant odour. The creature stood for a while and looked at the hunters, and then the wind drove the smoke and sparks in his face. Then he was gone, and they could hear how he broke branches and bushes in the forest. In the morning, the hunters found and measured his tracks.

Malik Saidov also gave us information he received from his late father. The father was hunting near Obigarm in the mountains, where he killed a goat. He did not have time to go down, so spent the night in a cave. He skinned the goat and started to fry the meat. The cave was located on a steep slope, about 100 m down from the cliff top. Suddenly stones started falling down. The father looked up and saw a woman hanging upside down in the cave entrance. She had large breasts and was covered with dark hair. When she saw the man looking at her, she pulled herself up, but then fell down again. The second time Saidov's father shot her between her breasts. She rolled down, and at the bottom she shouted in such a terrible voice, that even the mountains answered. Saidov's father did not sleep until morning, waiting for other creatures to appear. When the sun rose, he went down the cliff and found the corpse of the dead creature. It was a woman, but her whole body was covered with hair, even the fingers and toes; her nails were longer than a man; her eyes sat very close together. "The coat was very luxuriant and long, so my father wanted to skin the creature, but the woman emanated such a stench, he did not do it and left. My father later warned me about hunting like that and shooting something like that. Father also told me about something that happened to his father, who was a lifelong hunter, and a tall and strong man. Grandfather also lives in the Obigarm region. One night he was returning from the mill, carrying flour on a donkey, when a *hum* grabbed him close to a ravine. They started to fight. The donkey came home alone to his wife with the sack of flour. But two hours later my grandfather came home too. He had strangled the *hum* and left the dead body on the road. The next morning no one dared to go down the road past the dead *hum*, so my grandfather had to go back to the place and bury the dead creature." Moving west, we find information about the *hum* in an area almost all the way up to the city of Dushanbe. The hunters usually see him in the mountains. He looks like a man, is covered with hair and emits a very strong odour. The story goes that in 1937, near the villages Chukmayran, Tundara, and Bedak a *hum* was living in a ravine. It looked like a human creature but was covered with yellowish hair, almost like a camel. They say that it is still found there now. Tajik Abdugaffor Saburov, 46 years old, a disabled World War II veteran, was hunting in the area in the fall of 1948 around Ramit, near the village of Canas. At a distance of 500–600 metres he saw a huge humanoid creature, covered with hair, with large hands and feet. The hair was longest on his shoulders, thighs and breasts. It gave off a powerful stench.

Further south, near Faizobod, residents say that there were *guls* living in many places in caves in the mountains. They would wander out at night and attack lonely travellers. In 1960, residents of the village Kanaski-Poyon said that in the winter a herd of "wild men" would roam the vicinity of their village. Once in a very cold winter, three of them came into the villages and ate the village garbage in full view of the residents. The local residents also told about a village called Dar-Almast on the Kanas River. The word *almast* or *almasty* we already know. The villagers also claimed that the guls lived along the river. Shepherds from Sard and Miyona allegedly often see the guls in abandoned orchards. When describing the wild men, the residents of the upper reaches of the Kofarnihon not only talked about hairiness and large size, but also about very large eyes (perhaps deep eye sockets?). And the impression of four-fingered hands, which could be explained by having thumbs very close to the palm of the hands.

If we follow the Vakhsh and Kofarnihon Rivers in the south-western part of Tajikistan, we find information about "wild people" here and there. In these places, we are told, hunters sometimes kill a *hum* (*yavana*, *almast*). Some years ago, near the village of Vakhsh, people would die in mysterious circumstances. Sometime they would disappear altogether or be found dead with broken hands and cracked heads. Finally two border guards were lucky, and they caught a great woman with long breasts. At first they followed her into a cave above the river, and caught two of her male young as well. All three were covered with sparse grey-yellow hair, and they screamed loudly when they were tied up. Even far into the Hissar Mountains I have found stories about "wild men". V.A. Shake has collected a number of interesting observations in the valley and around the Kofarnihon and the Babatag Range, for example the find of a pile of stacked tortoises. The following story is one collected by him.

Kinji Olimova, 45 years old. Many years ago, as a young man he met a *maimon*. The meeting took place in a forest in the Babatag mountains. The maimon was short, about one and a half metres, overgrown with small sparse hair of a blonde colour (this could indicate that the maimon is a half-grown wild man). The upper part of his face was hidden by strands of the same blond hair. Olimova was especially struck by his red lips. When they met, the maimon lifted his hands, shook them and become strongly agitated.

According to a Tajik witness, in 1944 in a village to the north of the city of Dushanbe, in the Varzob valley, hairy "wild men" (*hum-yavoi*) suddenly appeared and started greedily eating the ripe mulberries. The man explained that in that year the crop of mulberries in the mountain above the village was a complete failure, so the wild people had nothing to eat, and went down to the village. The area situated between the Karatag and Shirkent Rivers, and the Karatag Valley stretching to the mountains to the Mura Pass leading to Iskanderkul Lake, was once, according to the nomadic Uzbek people, inaccessible to people. A Tajik villager from Hakimi said that the Pashmi Valley has a rich walnut forest and water, but had long been devoid of people because "the hairy wild men from the mountains would push boulders down the mountain and throw stones and force people to leave these places." According to G.K. Sinyavski there was a rumour that most villagers from Hakimi practiced a mysterious craft. They smelted iron ore from the local area, and at the same time melted out the fat from killed wild men for the production of precious medicinal drugs. We are very interested to know if there is any truth to this.

There is a considerable number of sightings from the slopes of the Hissar Range. For example Abdullah Jan, a resident in the centre district in Hissar, one evening found a "wild man" *(ëboi-odes)* lying in a pit they make for overnight stays in the mountains. It was 2.5 m in height and covered with black fur. Abdullah was so afraid of the wild man that he could not shoot. The wild man was frightened as well and ran away. It is also interesting to note that on the southern slope of the Hissar Range there is a gorge called Almasty Darya.

In the middle reaches of the river is a gorge called Duzah-gift, which can be translated as the the Gorge of Hell. G.K. Sinyavski has pointed out this gorge and the surrounding area as the habitat of devas. They are the raw material for the drugs mentioned above. Interestingly, we have a completely independent report from the same area, from B.M. Zdorikov. His guide Yusuf Yabukov told him that in the remote Duzah Dar canyon a naked hairy person sat by his fire.

In July 1961 I made a trip to the Karatag Valley (together with zoologists S.A. Side Aliyev and A.I. Kazakov). High in the mountains near the Kaltakol Pass, we met a shepherd called Nuchaev (Tajik, 57 years old, Lenin collective farm, district of

Hissar). I asked him whether it was true that in the old days this area was home to the wild men. Nuchaev said yes, and that they now can be found in the Duzah-gift gorge, which is in the middle of a lot of inaccessible lateral cracks in the rocks; there may be other unknown animals there as well. To see the *odes-yavoi* in Duzah-gift it is necessary to wait in some of the densely overgrown rock cracks. Surveillance should be conducted in the morning and in the evening, as the odes-yavoi sleep somewhere in the shadows during the hot daylight hours. If you want to catch the odes-yavoi, he added, it is necessary to send a hundred armed men to surround the valley.

My journey was too short to implement these recommendations. But of course I had to visit the Duzah-gift gorge. An experienced guide led our group to the upper canyon (south of the village of Labidzhay). It descends steeply towards the east for 10–12 km to the exit to the Karatag Valley. The Duzah-gift gorge is a majestic cut in the rough stone. The slopes are covered in unusually lush vegetation. This is not hell, but truly a paradise for wildlife: it is filled with plums, apples, wild rose, walnut, almond, hawthorn, blackberry, and many wild rhubarbs. The river at the bottom of the gorge is supposed to be full of fish (possibly trout?). We also found tracks of wild boar and bear feces. According to our guide, in the upper part of the gorge there are bears, wild boar, lynx, wolf, marten, fox, badger and porcupine. The descent down to the bottom of the gorge was extremely difficult, and we had to go back up. The vegetation is so dense, it would be virtually impossible to get through without special means.

If we go upstream on the Karatag, we find ourselves in the middle part which is enclosed on both sides by steep cliffs. In the spring of 1960, V.A. Shake and A.I. Kazakov saw and photographed tracks that reminded them of the Shipton footprints. Unfortunately the picture of the tracks on the dusty pathway is far from clear. Shake believes that in the area at the time there were four individuals: a male with large footprints, a female with smaller footprints, and two young with very small footprints. A shepherd named Nosyma Hidyrova from the Karatag village has also seen footprints that he thinks belong to the almasty in the area.

On the right side of the Karatag valley, slightly below Hakimi village, lies the Sarbin Valley. During several overnight stays in the Sarbin Valley Shake and Kazakov clearly heard several strange cries from unknown creatures. They distinguish two types of these cries. The cries begin at sunset, die down during the night and resume at dawn. The screams were coming from 2–4 different sources that moved around. When V.A. Hodunova tried to imitate the cries, they got a clear response (there are no echoes in this area). A local shepherd told them that he once saw the dismembered corpse of a mountain goat with no head in the snow on top of the Zagargo mountain. There were tracks of a barefoot man all around it. The area does not seem to be a particularly favourable environment for such a creature. The vegetation is sparse and the area is not isolated from the local people. However, the presence of bear droppings and various rodents indicate that at least there would be enough food.

One of the oldest local inhabitants reports that he remembers how, shortly after the Revolution, one of the inhabitants in Karatag brought a *maymunah* down from the mountains. After some time it ran away and hid in one of the caves the villagers had dug when they were looking for clay. One day he almost fainted with fright when the clay man jumped on his back. After some time he disappeared.

Ziadullo Rakhimov (Tajik, 34 years old, worker, resident of the village Shahrinau) said that in 1951 or 1953 at the bridge near Shahrinau (a bridge on the road Dushanbe -Termez), he saw a group of people gathered around the famous Russian hunter Nasadskogo, who was returning with his dog from a hunt in the mountains. Coming closer, he saw that the people were looking at something Nasadskogo had in a bag. It was the size of a human baby a few weeks old. It was a male baby, it had grey eyes, and it blinked frequently, but did not move. The baby was covered with sparse reddish or brown hair, except hands, feet and head (the hair on the head was quite short). It looked very much like a human apart from the hair, and the toes seemed long. It did not look like any monkey they had ever seen. The hunter told them that the creature was caught when four dogs, that were trained to hunt wild boar, surrounded a large nest under a bush and started to bark. The wild man baby was fed milk and raw meat. Local elders persuaded the hunter to let the wild man go after some time, as it was believed to bring bad luck. The only thing we know for certain is that he disappeared after some time.

Several residents from the village of Karatag (Sharofat Saidov, Gafar Subkhankulov and others) have told us that in the winter of 1959 a hunter from their village called Ziadullo brought a maymunah from the upper Karatag river. Many residents of the village came to see it. Although similar, there are some differences between the *maymunah* or *maimon,* and the *odes-yavoi, almast,* and *almasty* that is the "wild man". The lower height; a light-coloured coat; less hair on the scalp, and the chest, and almost absent on the buttocks; more frequent use of four-legged locomotion. It is also alleged that the maimon is mentally different from the wild man, i.e. smarter than him. Someone allegedly noticed a maimon with rudiments of a tail. It seems the maimon does not descend into the Karatag Valley, nor lives high in the mountains, but remains primarily on forested slopes, where it is sometimes found in artificially dug pits and caves.

In July 1960, V.A. Shake and N.M. Sadullayev were entering the Yangoklik-gift gorge on the left side of the Karatag Valley when they heard strange sounds like a baby and a young goat crying at the same time. They did not pay much attention to it. Only when they talked to a shepherd called Nosyma Hidyrova in the next valley were they told that they had heard the cry of a baby maimon. In the summer of 1961 we visited the same gorge, and could see that the slopes of the gorge are very richly covered in vegetation, and it seems to be home to a substantial number of large animals such as bears and wild boars.

Just above the left side of the valley, there is another valley, the Karatag Timur. In the middle of that we find the Timur-Kul Lake. We saw a lot of animal tracks on the trail along the shore of the lake. We found bear tracks that were 26 cm long, and a few tracks of a bare human foot, about 30 cm in length and 15 cm in width. The best track was from a right foot. The big toe about 1.5 times bigger than the second. The heel was 5–7 mm narrower than the foot, but rounded. Unfortunately the ground was so dusty and the tracks so unclear we could not make plaster casts or take good photographs.

Even higher on the left side of the valley is the Karatag Payron river valley, at the top of which is a fairly large dammed lake called Lake Payron. There are a number of reports of wild men from this area. Some people told me there is a family of wild men, a male, a female and one young, living near Lake Payron. They are even called *odes-obi*, which means water people. A hunter called Khikmatullo from Karatag told us that when hunting in the summer of 1958, he was dozing by the fire at night at the shores of the lake when he suddenly saw an odes-obi coming out of the water. He was so frightened that he couldn't reach for his gun. The odes-obi just stood knee-deep in the water, and then disappeared in the dark.

It turned out that the entire left bank of the lake is completely inaccessible to people, as well as the entire left side of the turbulent river for 8–10 km. Quite a large area of mountainous terrain is at the top of the ridge, and there are no passes at the bottom. There are no crossings on the river, but lots of narrow rapids. If a family of wild men really lived in this natural reserve, they can only appear on the right side in summer, when they can swim across the lake at the bottom of the narrows (50 m). At the top of this landscape we find large alpine meadows, ideal pasture for mountain goats. Below this is an area of stony grassy slopes teeming with marmot burrows, and below these is a vast floodplain with a rich plant life. A variety of trees, lots of nuts, plum, wild apple, wild and thick grass with fantastically bright colours. All of this is exposed to the south. A family of wild men could find food here throughout the year. In early spring of 1961 Kazakov and Shake discovered traces of a bare human foot at the site of the river run-off from the lake. In the summer months, the wild men could not only move to the right side of the lake, they could go down the valley, even try to scare people away from their place of residence, as has often been described in the reports from other geographical areas.

This would explain the following two personal observations done by A.I. Kazakov and V.A. Hodunova in this valley. On 1 May 1960, around 5 p.m., during a break, they heard a loud cry of surprise from the mountain above. They could see an animal walking by on two legs. It was difficult to see its outline in the snow, but it looked black. "For a few moments it stood motionless, then dropped down on all fours and quickly galloped to the top. There it got back on its hind legs, crossed the mountain and disappeared on the other side of the ridge." The whole event lasted for 5–6 minutes. Another incident occurred during an overnight stay from 28 to 29 October

1960, also on the right bank of the river valley. Around 3 a.m. Shake woke up with a sense of alarm and woke Kazakov. When they looked outside, their fire was still smouldering. Suddenly the tent leaned over, “and at that moment we heard footsteps. The animal undoubtedly passed near the fire, as the firewood that was stacked two feet from the fire crunched under its feet. After this the animal proceeded onto the stone scree, where we could hear stones rolling down for about 1.5 minutes. Then it headed back towards the tent. I rushed to unfasten the entrance. At this point the animal was at the back corner of the tent, and made a terrible cry like I have never heard before. Then it ran to the river and disappeared upstream very quickly. About 10 minutes later, when I was looking at the area, it cried far up ahead, somewhere near the dammed lake. We are certain it was not a bear. It sounded human, and the tracks were about as large as size 43 human shoes.”

The Iskanderkul Lake is also the centre for a number of reports of “wild men”. According to Hodunova, the villagers say that there are several families of wild people in the area. According to a shepherd called Kinji Olimova, one of his friends has seen a maimon in the area around the pass from Dukdon. According to other villagers, the wild men can be found in most of the area.

There are only a few interesting data from the Zeravshan Range. Amon Valiev Daheus from the village of Zeravshan says he has seen the creature. It is covered with hair and it is taller than a person. It is very afraid of people. Other residents have also seen hairy wild men near Zeravshan. In 1933 the geologist V.I. Sobolewski talked to an old man in the Rova-sung Pass. He was helping and guiding the geological expedition. Here on the snowy hill he had once, when coming down from the ridge, seen stones falling around him. Suddenly one of them hit his shoulder. When he turned, he could see the top of two monkeys, a male and female. He was using a bow for hunting, so he shot the male. The female gave a loud cry and disappeared. The hunter climbed up and found the dead monkey. It was covered with grey-yellow fur. The hunter dragged it down the snowfield, put it on his donkey and rode home. On the way, he met the secretary of the nearest village. He showed his prey to the secretary, and shared his doubts that had started along the way. What if he had not killed a monkey, but a human? The secretary told him, if you have killed a man, you will go to court. If you have killed a monkey, the Russians will not let you rest, but will require more monkeys. The hunter then buried the body somewhere near the ridge under the rocks.

Finally M.H. Khaydarov, who is a teacher at the Uzbek State University, has told us about some memories from his grandmother, from around 1866–1867. There was a big gathering in the streets of Samarkand, and there was also a wild man. He was hairy, but had a human face, and sparkling little eyes. His hands were tied behind his back.

The next set of information is from the Kashka-Darya and Surkhan-Darya region in the Uzbek SSR. Some very interesting records have been collected by G.Y. Snegirev when he lived in the village of Urgut in 1960. A shepherd named Usman Ruzibov and his brother Aktham would often chase the sheep that wander through the valley. There are no bears here. The humanoid hairy beast they call *aikh*, but according to Professor N.I. Leonov they also use the word *yavna*. The aikh supposedly lives in the mountains, and in the evening it will sometimes come down to the alpine meadows to gather food, and then return to the mountain before dawn. It will dig up tulip bulbs (*lala*), the roots of rhubarb (*chukrov*), and eat from the high wheat-fields as well as other herbs. It will also eat mice, which it catches and strangles with its hands. And it will also take carrion. Usman Ruzibov has met the creature in the pass at night. It was very close to him, coming up from the meadows. In the morning Usman found footprints in the snow. They were about 30–35 cm long with a big toe sitting back from the front of the foot.

According to the records of G.Y. Snegirev, the Tajiks believe that the aikh normally does not touch a person, but it can kill if it wants to. An old shepherd called Saipem, from the village of Urgut, was going home from the mountain with an injured leg two or three years ago. When he turned around a rock, he came face to face with an aikh. He was very frightened, but the aikh did not touch him, and both went on peacefully. Another herder has told how a calf had been killed by an aikh. Five years ago, during the construction of a road through a mountain pass, road builders and drivers saw two aikh and one of them was killed. When asked whether the best way to catch or kill an aikh would be by using dogs, the people answered that the aikh moves very rapidly and climbs the rocks on all fours. When meeting with humans, an aikh will always be standing on its hind feet, making strange movements with his heads and hands. His hair will hang down over his eyes. Usman Ruzibov has also seen an aikh collecting juniper branches, and a hole that an aikh had dug in one of the summer meadows.

The aikh supposedly likes to sit on earth mounds in the mountains, so will sometimes leave imprints of his backside as well as footprints and handprints. According to the locals, the aikh is about 2 m in height and covered with reddish hair. The hair on the head is long, and the forehead is sloping. The aikh are becoming more and more rare in the mountains. They live in small groups, and do not go far. According to a group of shepherds, a male, female and two young came down to the creek to drink water. When they went back, the adults first took one baby and then returned for the second. When asked why tracks of the aikh are so rare, people explained that the aikh leave no trace on an alpine meadow (except for remains of the mice they have eaten), as the snow in the mountains, even with very little wind, quickly fills up any tracks. Nobody goes looking for the aikh as it is considered a bad omen to meet them.

Snegirev notes that among the people surveyed by them in the market in Urgut, the Uzbeks knew nothing about the aikh or a wild hairy man under any other name, and that all the information and stories comes from Tajik herdsmen accompanying the flocks of sheep in the mountains. Usman Ruzibov's grandmother told him that when she and her husband trekked to Afghanistan she saw an aikh for the first time in her life.

However, just because these stories are not familiar to the Uzbeks does not mean we are dealing with Tajik superstition. According to S.S. Semin, in 1943 in the villages around Sari-Assia (Surkhan-Darya region of the Uzbek Soviet Socialist Republic), one of the locals told him how he had met with beings similar to humans a long time ago. They were much taller, and seemed more human to him. There were two, a man and a woman. Their body was covered in hair. The man was holding a stick and the skin of an animal, and the woman was holding a child. When they saw the man, the human-like creatures hurried away into the mountains. Despite the fact that the ground was covered with snow, they were both naked. Some very interesting folk legends associated with the emergence of *Yavan-adam* ("wild man") in the Baisun and Shirabadskom areas were recorded by the well-known local historian and hunter I.F. Lomaevym.

We also have reports of "wild men" to the south, in the Turkmen SSR. Here is one of them. Hunter Chara Kulari, 70 years old, from the village of Polvart, says that at one time he was out hunting with a friend near the village. In the evening they lit a fire, and put the carcasses of a number of birds near the fire and covered them with sand. When it got dark, the hunters heard what sounded like a moan, and soon after saw a humanoid creature approaching them in the light of the fire. He had very long arms, a small nose, his broad and flat body was covered with thick hair like a bear. Yavan-adam stopped making sounds, began sniffing the air and went straight to the place where the dead birds were covered with sand. Then he began to dig in the sand with his hands, grabbed one of the carcasses and started greedily to devour the flesh. Then he started digging in the sand again while making the moaning sound. The hunter Chara Kulari added, "Yavan-adam is afraid of fire. But we cannot shoot and kill the wild men, since they are not animals but people, our blood brothers, that live in the wild".

We will now return to the mountains of Tajikistan. In the north it has borders with the Turkestan Range, to the east with the Kirghiz SSR. Bobokalonov is an honoured worker of the old revolutionary movement, now a pensioner about 60 years old who lives in a village in the Shahrinau Hissar district. He has said that in the autumn of 1935, when he was with a detachment of the Red Army near the village of Match in the upper reaches of the Zeravshan, in one of the side valleys of the Turkestan Range at an altitude of 3000 m they came across a "wild man" (*odes-yavoi*). It was tearing at the grass and eating it. The wild man was driven out of the valley and caught. He was tall, about 180 cm. The face, neck and the palms of the hands were naked, and the rest of the body overgrown with black hair. The hair on the head was long enough to fall down over his ears. The forehead was sloping and the eyes were big. He was then allegedly sent to the museum in Dushanbe.
An interesting report comes from G.K. Sinyavsky. The incident took place around 1916–1919. A group of Austrian prisoners of war were in the city of Fergana. In a

pile of bones in one of the mountain passes leading from Fergana, they found the bones of a strange ape. The bones were photographed, but the fate of the images is unknown. According to Sinyavsky, this could have happened in one of the following passes: Tengizbay, Caique, Kaindy or Kichik-Alay. In the Alay valley he made a series of recordings of the observations of the population and cases of capture and killing of the gul-biyavan. One interesting story was told by two responsible and credible workers, the chairman of the village council in Kara-Kul and the collective farm chairman, about an event that took place not far from Dar Barrow-out. A man named Homar Chalobaev saw a gul-biyavan in a pasture. He ran to a nearby village, seized a gun, brought two men with him, and managed to shoot the creature. However, when seeing how much it looked like a man, he thought he would be punished for shooting a human and buried the body in the ground. In 1958 the author of these lines managed to track down one of the participants in the hunt, Ishanguly Gafarov, and the brother of the late Homar Chalobaev, but none of them would talk about the event.

Anthropologist Yu Rychkov also gave an oral report about an episode where people used a lasso to catch a live gul-biyavan in the Alay valley in 1957.

Kyrgyzstan

And now we come to the Soviet Tien Shan (we have already covered the Chinese Tien Shan above). From this area we have some information, but all fragmentary and random. So far no systematic scientific and biological work has been done in the region. There are also serious problems in the form of common superstitions about these beings that are being supported and exacerbated by Muslim mullahs.

In spite of the national homogeneity of the population in Tien Shan (Kyrgyzstan) we find, as everywhere else, a great variety of names for this human/wild animal. Some of the common names are considered to be of clearly mythological creatures, such as the *gesu tyrmak* which ethnographers and folklorists have shown are fairy-tale female werewolves with copper or gold claws. Further research has shown that the term is not just used in the mythological sense in the area. Ismail Shiraliyev, 73 years old, forestry guard near Lake Sary-Chelek, explained that the name gesu tyrmak ("metal claws", "copper nails") is a figurative way of expressing the very large finger strength of these creatures. They can hold on to something with a lot of strength. It is possible that such is indeed the original meaning of the term, and that the tale of the copper and gold claw is a later reinterpretation. Gigha Babataeva, 72 years old, mother of the forestry shepherd in the region of the Kara-Suu Lake, said that when a hunter killed a gesu tyrmak "the hands were completely the same as men".

Shatman Borukulov, 75, a farmer from the village of Abdykalykov, said that there are both male and female gesu tyrmak, and that they are similar to humans (including their hands). Many years ago a hunter named Kulanday allegedly shot a gesu tyrmak male, female and a young female in the mountains below the Chatkal Mountains. *

The Kyrgyz generally consider the gesu tyrmak to be animals, but with much in common with humans. They live mainly on the big mountains, have a dark brown colour and strong claws on their fingers. A large number of them can be found in the high mountains on the road between Kokand and Kashgar. The gesu tyrmak live in pits or caves lined with stones. However, the term gesu tyrmak is not the only or indeed the most common term for wild humanoid animals in the Tien Shan. A more common term is *kishi kiik* ("wild man"), but we have also found the name *albarsty* although this is closely entwined with beliefs in ghosts and other beings.

A reader wrote to the newspaper *Komsomolskaya Pravda* that in the winter of 1941–1942 there were rumours in Jambul that people had seen one in one of the surrounding areas. It was a humanoid, all covered in hair, that moved on two legs. Another reader of the newspaper wrote that during ten years living in Kyrgyzstan, he had heard of the "mountain man", including stories about shepherds that saw him twice in October and November 1957. The reader adds: "It is significant that in all cases there is one common characteristic – the 'man' is covered with hair from head to toe". Here is another letter to the editor of the same newspaper: during 1945–1950 the writer often visited the area around the Jalalabad region on the slopes of the Fergana and Chatkal Ranges. During this time he heard from an old Kyrgyz hunter about *zhestonschuki* – terrible hairy people living in caves high in the mountains. Sometimes they would attack lone travellers. An old hunter in the Achinsk district even claimed that he and two comrades were attacked by zhestonschuki in the Baubash-Ata Mountains, but he managed to escape his pursuers.

The well known geographer Professor RD ** heard abundant tales of gesu tyrmak in Tien Shan. This animal is also called *zhelmoguz* (the eater or glutton). The professor wrote two reports about gesu tyrmak being killed. One hunter talked about seeing three humanoid beasts from afar. They were collecting bundles of *kischlichnaya* (a kind of rhubarb). The hunter shot and killed one of them, but the two others dragged it with them when they fled. The hunter and his horse had a nervous breakdown from hearing its dying cry. And finally we have some interesting biological information recorded by R.D. Zabirova. "People who have seen the gesu tyrmak says that his hair looks like the grey wool of a camel. His body is like a 17–18 year old boy. The face is very similar to the face of a monkey. The nails are long. It usually walks on two legs."

Many stories come from the foothills of the Fergana Range in the Oktyabrsky district of the Jalalabad area. The people living in the thick walnut forests near Arslanbob say that in ancient times, when the first settlers came to the area, there were hairy animal-

* *Translator's note: A seemingly meaningless section has been removed here.*

** *Translator's note: Possibly another case of Porshnev hiding someone's identity.*

like men living in caves. In our times, they say, there are a few old people who can remember how people were afraid to go near a special narrow ravine, because you would hear terrible screams from the cliffs and stones would start flying towards you. But even fifty years ago, as reported by Hidoyathon Nurmatova, a shepherd ran in to Arslanbob and said that in the "cursed valley", where the *yavvon-adamlay* ("wild man") was living, he had seen an animal standing on two feet with a big club in his hands. He was tall, but short-legged, not wearing any clothes, and his body was completely covered with thick dark hair. It could not be a bear because of the shape of the head and especially the hairless face. It was a bestial creature akin to a person. The forehead was low and sloping, and the eyes were small and narrow. The shepherd said he threw a lamp with lit oil at the creature and then ran, while the creature screamed wildly.

In 1947 a scientist named Y. Vinogradov heard stories that in the area around a state farm in Kyzyl-Uygur there was an *adam-ide* ("man-bear") living among the rocks. Some of the residents said that they had seen the adam-ide from a distance, but not during the day or in the evening. Sometimes you would hear it screaming in the afternoon. According to descriptions, the adam-ide has no clothes, but is covered with hair from head to toe. It is built like a normal person. Local people often left food out for adam-ide near the road leading up into the mountains. But the old Kyrgyz, when asked about the adam-ide, would turn the conversation to other topics.

Let us now turn to a study conducted from 1935 to 1958 by Professor A.A. Mashkovtsev in one particular area of the Tien Shan. During field work, and then through correspondence with the local collective farm workers, Prof. Mashkovtsev found that in the desert as well as in the rest of the valleys of the north-western slope of the Kyrgyz Range, from the western ridge of the valley to the east of the Dzharlykaindy Issykaty valley, i.e. in an area about 120 km in length and about 55–60 km wide, he could find information about meetings with "wild people", not only in the distant past but also the present. According to the stories of the Kyrgyz many of their ancestors who settled in these places were killed with stones, thrown by hairy wild men that lived in the area. The main habitat was the Kyrgyz Range and the area just south of the Dzhumgoltau Range. According to the Kyrgyz hunters, the present day wild men were very small, and most of them were either dead or had gone somewhere farther to the south-east towards the Chinese border. They are still supposed to live in small numbers in the most desolate gorges of the Kyrgyz Range, in particular on the left slope of the Kara-Balta River valley and in some places on the Dzhumgoltau Range. Hunters from the village of Sosnovka warned that on the left slopes of the Kara-Balta Valley you could fear attacks from wild men. They would throw stones at people in the area.

Later A.M. Urazovsky, who has a horse breeding farm in Sosnovka, wrote in 1951 that in the upper reaches of the Kara-Balta, beneath the pass, he had seen tracks that resembled the tracks of a barefoot man, although with some differences. Hunters and shepherds only rarely see the tracks of wild men. At one time hunters shot and wounded a wild man, but did not catch him. They said he was "like a big monkey." In the upper left tributary of the Kara-Balta River, in the Abla Gorge, tracks of the wild man have been seen a few times both in the snow and in sand. "In 1951, in the Abla Gorge," writes Urazovsky, "I saw three marmots that had been dug up by the wild people. The stones were not thrown everywhere like when wolves and bears dig for marmots. They were neatly laid to one side. At the bottom of the holes were the tails, paws and fur of the marmots. Apparently each wild man would catch five or six marmots." According to Prof. A.A. Mashkovtsev, the special conditions in the area and the lack of pasture was the reason one could still find small relict groups of wild men.

If we look at a map, we see that the river Ak-Suu flows more or less parallel to the Kara-Balta River from the east. Apparently it was in a gorge of this river that the following story told by a Kyrgyz hunter named Kerimkul took place. It was about 60 years ago, one evening in the middle of winter when Kerimkul was roasting meat on a campfire in a cave situated in the cliffs of the Ak-Suu Canyon. Suddenly he saw a creature at the entrance to the cave. It was a female without clothes, but with the whole body covered with hair. She was hunched over, and one hand was supporting her breasts. She went carefully towards the fire and looked at the man, crouched down and took a piece of meat from the hunter's outstretched knife, and then retired. Kerimkul also claims that he had shot a gesu tyrmak twice. Once he cut off the hands of the creature and brought them to the villagers as a hunting trophy. The other time he shot a gesu tyrmak near the river, but the body fell in and disappeared.

In 1940 G.A. Petrov recorded a story of an old hunter from the nearby Suusamyr Range. He said that once when he was out hunting with others for red deer he stumbled on a "wild woman". She was caught with a lasso. She was of medium height and naked, but her body was covered with red hair. Apart from that she seemed only a little different from humans. She could not speak, but made a shrill cry. When the hunters tied her up, the wild woman behaved very violently and continued to scream and fight. In the end, the hunters decided it was too risky to try and get down from the mountain with the wild woman, so they released her. A report from the geologist M.A. Stronin concerns an incident from August 1948 in the same area of the eastern Suusamyr Range.

From the west of the Suusamyr Range comes a curious story from a Kyrgyz called Usen Tokhtasynov, about 45–50 years old, recorded in 1950 by the local historian P.T. Hemp. The incident occurred shortly before the Revolution. The grandfather of Tokhtasynov and his three sons and their families drove their herds and moved for the summer to a base camp in a completely deserted area. In the middle of the night they heard some screams unlike the voices of animals they knew. But it was enough to fire one shot in the air to make it stop. The shepherds had planted melons about a kilometre from the yurts. And then they suddenly disappeared. One evening Tokhtasynov's father came running into the tent shortly after they had heard some hideous screams. He was pale and shaken, and could only say "yavvoi-adam". Around midnight naked wild men overgrown with hair had come to the melons and started threatening people. They were growling and screaming. When one of them approached the hut, the watchman cried out, the stranger jumped to the side, they all heard a wild cry, and then the sound of many running feet. On a moonlit night three days later the three brothers, who were guarding the melons with rifles, were eyewitnesses to the next invasion. Hearing the sound of voices, they rushed out of the hut and saw two stunted, half-bent figures covered with hair. They turned toward the noise, and then started to run away with wild cries. At that moment, the brothers noticed that the smaller one was a female with pendulous breasts. The watchmen fired, and the female uttered a sharp high-pitched sound, fell, then rose again and limped away. While the men were reloading their guns the sounds died away. After examining their melon beds in the morning, the men noticed blood stains and tracks of bare feet. They were basically human-like, but with flat feet and big toes set to one side. When the grandfather heard the story, he at once told everybody to dismantle the tents and move the cattle away from the area.

Another interesting report comes from G.I. Tymoshenko, a resident of Tokmak. He and several of his friends, who were involved in transporting things by cart between the town and the city of Frunze, had seen a cage being transported though Tokmak sometime in the 1920s. It contained two human-animal or wild men overgrown with hair. They produced animal-like squeaks and bellows. However, it was impossible to check the story, as almost none of Tymoshenko's eyewitnesses remained among the population of the city.

Additional information about the Kyrgyz Range was collected in the summer of 1961 by members of a glacial expedition from the Leningrad State Pedagogical Institute and summarised by the geographer E.V. Maximov. They interviewed 20–25 Kyrgyz shepherds. All respondents are well aware of the name gesu tyrmak, which means a wild-haired man who used to live in the Kyrgyz Mountains. They had heard about the gesu tyrmak from their fathers and grandfathers. The majority of them believe that this creature now can be found in the Kyrgyz Alatau and surrounding areas, possible in Central Tien Shan. The gesu tyrmak supposedly still occur in inaccessible areas near the Chinese border. Five or six of them suggested the Kavaktau Mountain. One said that gesu tyrmak can be found in the area of the Ak-Suu Pass in the Kyrgyz Range and around Dzhumgole. They say that the gesu tyrmak are like people. There are men, women and children. They have no clothes and live in caves. They block the entrance with stones. They can run fast. Their height is like a human, the body is covered with hair and they walk on two legs. They also have long nails. They make a shrill cry that frightens people. About four or five of the interviewed people talked about gesu tyrmak like a real being without any fantastic features. The rest talk about fantastic features like having iron and copper nails, just one foot or drinking human blood.

In 1942 loggers are supposed to have seen a wild woman with a child near the confluence of the Chon Torah and the Sokuluk Ashu-Tor Rivers. Members of the expedition also talked to Dzakshena Abdiev, the shepherd of the Karl Marx collective farm. He said that in 1942 he was 10 years old and lived with his parents in the Char area. Below the confluence of the rivers was a dense forest of mountain ash and junipers. One morning the boy went into the wood, and saw a completely naked hairy man at a distance of 15–20 m. The hair was yellowish-grey in colour. When the man saw the boy, he bared his teeth. The boy ran away. No one believed his story, but in the night they heard strange cries close to the camp. The next day the boy went out into the forest with three adult loggers. At about the same place where he had seen the wild man, all four of them saw a wild haired woman sitting on a stump. She was nursing a small child, and there was another child standing nearby. The wild woman did not notice the loggers coming up, but they felt it best to leave. On the third day, they decided to organise a raid, but with no result. When he was interviewed, Abdiev was presented with photos of Shipton's tracks, as well as images of heads of chimpanzees, Pithecanthropus and Neanderthal. Abdiev noted that the track of the wild man is narrower than the photo. His head looked like a Neanderthal. When asked about the shape of the head, Abdiev said it was pointed.

Based on the survey E.V. Maximov stated that he thinks we are talking about a real live or possibly extinct unknown animal or wild man.

Now we come to the Chatkal ranges of the Tien Shan. Here we have information provided by two Russians, but collected with an ethnographic point of view. It is important to compare the information of the indigenous population with the observations of travellers of other nationalities. A hunter named V.S. Bozhenov said that he once travelled with a group of friends for 700 kilometres to the west. They wanted to fish in Ala-Buka - the foothills of the Chatkal Range. When they got to the mountains, they could see a Bigfoot at a considerable distance. They watched it through binoculars from 9 a.m. to 2 p.m. Then it rushed down a steep rocky slope to a cave. Bozhenov writes that he tried to get to the cave, but it was necessary to walk in a large circle to get there, and the fog started to come, it began to snow, and not far from the caves the man had to return because of the weather.

At the foot of the Chatkal Range lies the perfect mountain lake Sary-Chelek. In July 1948 A.P. Agafonov was working here as an engineer and geologist. One night in a yurt he heard the following story from a more than eighty year old Kyrgyz shepherd named Madyer. According to his story, his grandfather once returned to the festival with his young wife. They were sleeping somewhere to the south-east of Sary-Chelek, when he was awakened by her screams, and saw her being carried away in the arms of a huge red-haired humanoid ape. The grandfather caught the thief, killed him with his hunting knife and saved his wife. After hearing this story, Agafonov explained to the people present that apes are not found in Tien Shan. This only made the listeners smile, and the old Madyer did something unexpected. From a large chest he took a carved box and handed it to the geologist with the words "see for yourself". "In the box on a felt mat, there was an artfully dried hand covered with sparse, but quite long hair – up to one centimetre. The inner side of the hand was covered in short wool.

Judging from the size, it could only have come from a large humanoid animal... my surprise was so great I did not make any sketches or even a simple description of the item. Yet I remember the structure of the human hand perfectly. I was only confused by the brown hair on the back of the hand." The above information prompted the author of these lines to make a special search of the area in September 1959 with G.G. Petrov to find the dried hand. The old shepherd Madyer had died just three months prior to our arrival at the age of 98 years. He was a man of great fame among the local population. No one knew the ancient legends about the wild man of Tien Shan like he did, although the young people tend to treat it as old man's stories. We did not have much success, but I am sure we have found information on another casket and wild man hand.

This case is highly significant. There are plenty of other stories about hunters taking the hands of a wild man as hunting trophies and mascots. One reports says "the hunters killed a wild man. His hands and arms were yellow, his nails long and curved. The hunters cut off the hands and took them as talismans."

Finally we turn to the easternmost part of the Soviet Tien Shan. In August 1953 Colonel V.I. Ratseka was on a mountaineering expedition to Khan Tengri and Victory Peak. When they were moving up the Sari Jaz River valley towards the Engilchek Glacier, their guide Torment Maksutov, who is an expert on the area said, pointing to the mouth of the river and valley: "This is where the wild mountain people lived. They have been gone for a long time now. It is maybe 70–80 years ago. They all went to China." Maksutov added that the gorge is called Shilong.

Are all the "wild men" long gone from these places, then? On the contrary says geologist M.A. Stronin. "In August 1948, I carried out geological work in the Tien Shan in the eastern part of the Terskey Alatau. I once spent the night in one of the valleys adjacent to the Engilchek area, about 10–15 km below one of the tongues of the Engilchek Glacier. I arrived by night with two Kyrgyz people. Early in the morning I was awakened by the Kyrgyz, saying that someone was stealing horses. At a distance of about 500 m I could see someone was walking around the horses. I quickly went to the horses with one of the Kyrgyz. Now I could clearly see someone in among the horses. It was walking on two legs, but tried to huddle down behind the horses. The person's hands were long, much longer than in humans. First I thought this is someone with a robe with long sleeves, so I cried out in Kyrgyz, 'Why are you stealing!' When I shouted, the creature stopped and look round, there was a muffled guttural sound like the cry of mountain goats; and then the being moved away from the horses, first quietly, but then it turned and ran. In total I saw him for 7–10 minutes. I was not blinded by the sun, and I saw it very well."

"When it ran, I first thought that it was a bear, but a very slim one. It was running up a slope of about 30° very rapidly, using the forelegs for help like a horse galloping. It ran a little sideways as well. Being a hunter I was ready to shoot, but I could also see it was not a bear. The bear has a snout, but this had a much more rounded face. The fur was short and dark brown in colour, without the yellowish tint you see in a bear. I have hunted for a long time and seen every beast there is, but never anything like this. It was very special. A little man, but at the same time a beast. I was not sure whether it was man or beast, so I did not dare shoot. In the end it disappeared from view on the slope toward the glacier."

"Unfortunately I did not look for footprints, as I did not consider them important. Our horses were terrified. The Kyrgyz guides called it *kiik-kishi* ('wild man'), and absolutely refused to follow me any further. A little later in the same year, when we were in the Kavak Tau area, one of the guides who went out in the morning mist for the horses, came running back in terror yelling 'kiik-kishi'. In general the Kyrgyz do not graze sheep in areas where there is evidence of kiik-kishi. They are particularly reluctant to graze their herds on the top border of alpine meadows adjacent to the glacier."

Kazakhstan

The available information relating to the Bigfoot in Kazakhstan relates primarily to areas adjacent to the Tien Shan. Some very interesting stories about the "wild men" of the mountains of the south of Kazakhstan has been sent by an observer named Tomirali Borybaev from the Aksu Dzhebaglinskogo Reserve, located in the north-western spurs of the Talas Alatau. According to the stories of the Kazakhs there were wild men living in the mountains of the Dzhebaglinskogo Reserve. They were quite wild, with dense short fur very similar to dromedary fur. They had no clothes, could not speak, and ate meat and different fruits and roots. They had a ferocious temper and avoided contact with humans. Regarding a meeting with the wild man, he had heard about one a long time ago in 1870–80. A close friend of Borybaev's father called Samal-Mergen, who died a very old man in the 1920s, had been out hunting in the mountains. On one of the slopes he saw a creature that looked like a man. It was bent over, but straightened suddenly. The hunter realised that instead of clothes, his body was covered with thick short fur the same colour as camel or dromedary fur (greyish-fawn). The wild man was rather tall and muscular. He pulled some small plants from the ground, cleaned the dirt off the roots, and ate them. The hunter decided to shoot him in the leg. The wounded wild man yelled like a human, sat down, looked at his leg and licked it. For a while he sat and whined, but then he stood up and, limping badly, went towards the rocks and disappeared behind the slope. The hunter followed a little while later, and did find a trail of blood, but in the end lost it among the rocks.

At the moment the amount of information from this area is very much poorer than for instance from Pamir and Tien Shan. It is very fragmented and untested, so it seems premature to form any analysis or conclusion. The picture can only change in the future.

Altai, Sayan

Now we will take a brief look on the data from another territory which seems very promising for further research. A mountainous area covering the Altai and Sayan, including the territory of Khakassia, Tuva and the Buryat Republic.

Courtesy of J.F. Tsikunova we have the recorded memories of a very old man, the late A. Mogilev, dating back to his youth, i.e. almost to the 1830s. We are dealing with the border areas of modern day east Kazakhstan and the Altai Territory. Mogilev worked at a farm, located in the mountains far to the east of the city of Leninogorsk. The mountains are covered with snow, the slopes are wooded, and there is almost no population. At one time they heard a loud scream from outside, and caught an extraordinary man. He moved on two legs, but was covered in long hair. He was brought to the village and held until evening. According to Mogilev, he sat all day and wept, and shouted something unintelligible. When he was released in the evening, he fled to the mountains. They often heard the same cries near the village; sometimes they would put out bread and meat. In the morning it would always be gone and there would be tracks of the wild man. Mogilev also told about a local teacher, who was hunting in the mountains when he found the den of a flock of "wild men" in a cave. There were no adults, but several young in the cave. While the hunter was studying the cave, the adults returned and attacked him, but he drove them away by firing his gun into the air. When he departed, they rushed to their young, and the hunter managed to leave unharmed.

Based on this report it is clear that the Altai Mountains were a Bigfoot breeding ground more than a hundred years ago. There were sightings of young ones and pairs. But we do have some information that indicates that Bigfoot can be found even today in places like the Abakan Range. When the Estonian J. Nerman arrived in Abakan in 1938 to study the possibilities of building a resort around the hot sulphur springs in the Khakassia Mountains, he heard stories from many people about mysterious encounters in the mountains. The most recent took place only two weeks before his arrival. A wild man was also once captured in the mountains. "The local people thought it was simply a feral man that had been left in the mountains as a small child. They say that hunters and forest rangers have seen a hairy man at a distance in the mountains, and that nobody can get close to him. Local officials had ordered the people to catch this forest dweller alive. It took a long time before they could catch him with ropes and tie him down. When he was caught, he was biting, growling and screaming, but did not make any intelligible words. The wild man was carried to Abakan on a stretcher. This was placed in a special iron cage and shown to the public for a week. A few people did say that they went to see the monster. It was said that when people approached the cage, the wild man would growl and shake the iron bars so violently the whole prison would shake. He was fed with raw meat and fish. The wild man especially liked the fish, tearing it into pieces and eating it. They said he was tall and strong, and covered in thick fur. When I asked the people where the wild man was, they told me he had been taken away somewhere, while others added he was taken away in a special car, probably to Moscow." In 1955 the artist B. Zhutovki came to Abakan and heard a similar story about the capture of a shaggy man twenty years ago, but according to this story he had been sent to Delhi.

In August 1962, on the southern slope of Belukha (Katun) Glacier at an altitude of about 2500 m a mountain-tourist group found frozen excrement scattered over an area of about 2 square metres. It resembled human feces, but consisted largely of half-digested and undigested grass. Samples were brought to Moscow and examined.

We have a very interesting letter about a hunter who knew the area around Krasnoyarsk and Tuva very well, especially along the route from the town of Abakan to Kyzyl. He knew the tracks of native animals – birds and animals – very well. In 1952, on the crest of the Sayan Mountains, he found some completely unfamiliar tracks in the snow. They were nearly circular, around 25 centimetres in length, without any nail prints. It was clearly left by a hairy creature walking on two legs. And the hunter was very clear in stating that they were not bear tracks. After following the tracks for more than a kilometre, the hunter returned and found another track proceeding directly towards the starting place. When the local people heard about it, they said: "These are our ancestors, do not go near them, they will carry you away."

D.I. Popov from Irkutsk reports that in 1952, in a village in the Orlik Oka area, he had heard reports from local residents about meeting a very large humanoid creature with long arms and a white coat in the Sayan Mountains.

In 1954 three Azerbaijani loggers, Safar Saidov, Vaisala Akhmedov and Algeydar Ahundinov, were working in the eastern Sayan Mountains in the Buryat Republic. One day they went out to work on a mountain trail. "Suddenly some kind of monster stepped out from behind a huge block of stone. It was only 10–15 metres away. It had a body like a man covered with thick hair. He turned and ran very quickly away, holding something in his hands, and disappeared into the dark valley." When the local leader was told of the meeting, he was not surprised. He said that the wild man does not attack people, but lives in the mountains, where he catches wild animals and birds.

In general it would seem that the available data suggest that, at least in the historical past, this area has been an important area for the Bigfoot. One must also remember that this area is directly connected to the area with the largest amount of data on the Bigfoot – the Mongolian People's Republic.

Transbaikalia, Eastern Siberia

V. Konstantinov has received information from the Ayano-Maiksy region of the Khabarovsk Territory about how the Evenki hunters have told of rare meetings with the *bul* at the foot of the Dzhugdzur Range. It is a tall naked man overgrown with hair like a bear, a dark face and deep-set eyes. One of the stories concerns a case where hunters found a wounded bul, caught him, and tried to question him. But he only grinned and made loud cries that made their ears hurt. It is interesting that the observers claimed to have seen bul with a stick and a dry femur of an animal in his hands. These he used as percussion instruments.

To the north, in the mountains of Yakutia, we have a considerable number of stories about wild humanoids. Here they are called *chuchunya, kuchuna, meelkoeen, heedeka* or *abas*.

G.V. Ksenofontov, a prominent expert on Yakut folklore, has supplied us with a number of records. They do not consider the wild man a mythological creature, but a kind of primitive neighbour. The local people only meet the wild man on very rare occasions, and then he will flee immediately. In many stories it is emphasised that the wild man usually is alone, and that he can run with extreme speed. He is built like a normal man, but his body is covered with hair. It is possible that the information from this area is a mixture of folklore and actual reports, perhaps mixed in with dimly remembered stories of the Ainu people living in the southern part of the Sakhalin area. There are also legends claiming that when these hairy people are seen it will be the end of times, the end of the world, doomsday.

However, most of the stories sound more biological than mythological. The wild men whistle, frighten people and deer, and make a lot of noise in a very unpleasant, harsh and cracking voice. They are usually only seen during the summer. They live in burrows like bears, and live on raw meat from deer, possibly reindeer and even mice.

These stories can be supplemented by two cases collected by A.P. Okladnikov from the Lower Lena area. This is where one can find the *chuchunya* – a tribe of half animal - half humans that used to be common here in the north. Today you can still meet them occasionally. They looked quite extraordinary. They had no neck, so it looked like the head was fused to the body. They would sometimes appear at night and throw stones at sleeping people. They would hunt for deer around the river. A Yakut hunter named Makarov claimed that he found a cave inhabited by chuchunya on the right bank of the Lena River. Sometimes their lairs could be found on the left side of the Lena as well.

The geologist N.I. Gogina reports that during an expedition in the summer of 1960 two Evenki hunters and experts in the taiga, Yegor Pavlov and Vasily Fyodorov, were at first reluctant to answer when asked about the chuchunya, but then said that there are many wild men in their lands. They described the wild man as being very large, tall and hairy, and the footprints very large. "It is not a man, it is probably a Devil." The wild man can be found in the Verkhoyansk Range. It has been seen by reindeer-herders in the Kystatymskogo Area (Zhigansky District), as well as high on the ridges in the upper reaches of Dzhardzhan, Myangkyarya and Sobopol. They used to be more common, but now you don't see them as frequently. The wild men allegedly abducted children, but brought them back after some time. The children would then describe where and how they lived. Before the war there were more kidnappings, today they are rare. In one case they found the tracks of the thief. They looked like the tracks of "a big, big man".

So what do the local people actually believe? N.I. Gogina points out that the stories are very local. "In 1958 we asked all the herders about wild men, but no one knew anything about it. They were all living in the flat lands. But the herders from the Verkhoyansk area all told these stories." This is an indication of the stories possibly being fictional. Or put in another way, if it is a legend, it is not universally popular among the population in the province, and only confined to areas with specific geographical conditions. On the other hand it is interesting that, if it really is nothing but folklore, then the stories are very similar to the stories told by other people. Boris Leonov, another geologist, has also added: "The stories about the wild men in the Verkhoyansk area bear a striking similarity to other areas. The stories also describe the colour of the beast as being like the colour of a bear, and that it lives in a cave near the snowline."

According to one reporter, who worked several years in the north, in the area of the rivers Yana and Indigirka, the indigenous people told him that until recently the wild men had a meeting place on the Poloustnom Ridge, and that people in the Muurdah village had buried a dead chuchunya in the permafrost. The correspondent also says that "around the end of the 1920s, the authorities of Yakutia even gave permission to shoot a chuchunya, and that a Yakut friend of his kept a piece of skin of the slain creature."

The botanist V.A. Sheludyakova has reported that throughout the north of Yakutia people talk about "wild people" living in the mountains and other inaccessible places. The stories are so filled with details and come from so many credible witnesses that at one time Yatsik * was made specially responsible for figuring out what those wild people were - a product of superstition, or real people or beings living in uninhabited places. Several of the eyewitnesses who talked about their meetings with the *kuchuna* were literate people who would identify themselves as the local intelligentsia. Sheludyakova also tells an episode that occurred during her fieldwork in the region between the Indigirka and Alazeya. Some of her guides had killed a moose, and when they went out to clean and cut up the carcass in the morning, they came back and woke her up saying they had just seen a wild man come to the meat. He was short, dark, whistled, and threw stones at them.

In July 1959 the newspaper *Eder Communist*, the official organ of the Yakutsk regional committee of Komsomol, published an article by S. Cheremkina called "Do you know the heedeka?". The author cites an old collective farmer called P.A. Sleptsov from the Moma district at the Kolyma River. His story all relates to a single season, but the local people are generally well aware of the *heedeka* or "masters of

* *Translator's note: The text does not say who this person is.*

the mountain". Sleptsov described how in the autumn strange beings, not quite beasts, and not quite humans, would come at night to plague people. They would make noise and threw stones at the tents. He was lying in his own tent, but after some time he ran away and left his reindeer behind. Other people told about similar attacks. In the same year a relative went away to hunt sheep in the mountains. One night he had left out a carcass and the skin, when he suddenly realised they had disappeared. In the next instant something started throwing stones at him from several directions. When he fled from the attackers, they followed him, some running in front, and some behind him, still throwing stones. He tried to shoot at one of them, but could barely see a silhouette. The aunt of Sleptsov was once attacked by a group of seven tall, dark humanoid creatures, that made inhuman squeaky sounds.

At the moment we have too little data to say anything about the northern boundary of the distribution of the wild men. But we are probably talking about the Chukchi Peninsula and the Bering Strait. This is partly based on the quite abundant data on the same beings in America. They would just have to cross the Bering Strait.

The data from America has only recently been made available to us. But it, and other data collections, have shown that we are not the pioneers we thought. The only thing is that our scientific predecessors could not compare their local material with similar data from other geographic areas. It has turned out that in 1908–1912, a young Russian mineralogist by the name of P.L. Dravert collected (and published in a small way) a number of stories about hairy wild men from the lower reaches of the Lena River in Kuyusyur and Bulun in the Verkhoyansk district. Since 1925 Dravert, who became a prominent expert in meteorites, and the Yakutian agronomist D.I .Timofeev, have collected information about these wild men from several areas. In March 1929 they made a report on this subject to the Western Siberia (Omsk) Department of the Russian Geographical Society. On 26 April 1929 the newspaper *Independent Yakutia* also published an article by S. Potapov about the chuchunya, which basically confirmed that reputable Yakutian organisations have contributed to the investigation. Based on these and other data, Prof. Dravert published a big article in 1933 about "The wild people – muleny and chuchunya". The author calls for more effort in the study and protection of these "wild people", as he considers them of great scientific value.

However, this appeal went largely unheeded because of the simultaneous publication of an article by G. Ksenofontov: "Regarding the article by Prof. Dravert", which quite unfoundedly declared that "the muleny and the chuchunya are remnants of some ancient beliefs in the ancestors of the natives of the north, similar to the beliefs of the Ancient Greeks in the faun Pan." Ksenofontov noted that Pan traditionally is associated with herds of herbivores, the ability to make them "panic", persecution of women, and shouting in the mountains.

In his statement, Dravert distinguishes between two different ideas about "wild people". The first is based on the descriptions collected from the Yakut and the Tungus of a strong hairy humanoid lacking the ability to produce articulate speech. The second sees them as some kind of primitive people capable of using bow and arrows, metal knives and flints, and using clothes made of reindeer skin, possibly speaking a language incomprehensible to the natives. The two versions are of course mutually exclusive, but if we search through them, we end up with the following description of the chuchunya. The structure of the body and the general appearance are similar to humans, but they are taller, and are capable of running extremely fast. They are very hairy, even on the face, and they are physically strong. They will sometimes make a soft mooing sound while showing their teeth. In the mountains they will also make a shrill whistle before they attack. This sound can almost paralyse people and animals. They are sometimes armed with stones and sticks that they will throw at humans or their prey. They will occasionally attack humans to scare them, and they will steal their supplies. In the Tungus area they will steal deer, sometimes whole herds. They live in caves, either alone or in small groups of two or three individuals. There is no information regarding females or young.

Dravert and others have stressed that the Yakuts and especially the Tungus people are extremely reluctant to report their encounters with the wild people, because these meetings usually ended with them killing most of the latter, and they were afraid of being prosecuted for murder. It is rumoured that a considerable number of wild people were killed during the civil war in Yakutia. They have also been hunted, both before and after the Revolution. Their corpses are buried here and there, but always hidden. *

A final report serves to illustrate the character of some of the material for the "polar" chapter. This story was recorded in 1948 by ethnographer B.O. Long. It has the character of folklore, but with lots of detail. It is about three brothers who go out into the wilderness to fish, and meet up with a large hairy wild man, an *amook* sitting by a fire eating a roasted corpse of a dead man. The brothers decide to kill the wild man, cut off his head and bring it back to their father, who tells them that at one time there were plenty like the amook living in the area, but now they have all died at the hands of the Tungus people.

Returning to the research of Prof. Dravert, he ended up concluding that, since there was no mention of females or young in any of the stories he had recorded, it was reasonable to suggest that the chuchunya was leading a nomadic life, moving around in small groups, and only passing the area in early spring, and moving on for another as yet unknown destination. "The local people believe that the chuchunya come from the Chukchi district, and go back to it when the season is right."

* *Translator's note: At this point the author has inserted a note about a "polar" chapter to be written later about wild people reports from the northernmost parts of the area along with some notes that make absolutely no sense, so I have deleted them.*

There is a limited amount of information about the "snowman" on the Aleutian Islands. In a book by Ted Behnke we find references to stories about the *aydigadine* or "outsider" – a homeless terrible savage who prowls the mountains at night, sometimes close to the villages, where it will steal the food, and sometimes the children. This is a good link to the information about a substantially similar yeti on the west coast of Canada, in northern California and in other areas of America in Chapter 10.

Caucasus and Iran

A political map of the Caucasus as it was when Porshnev wrote

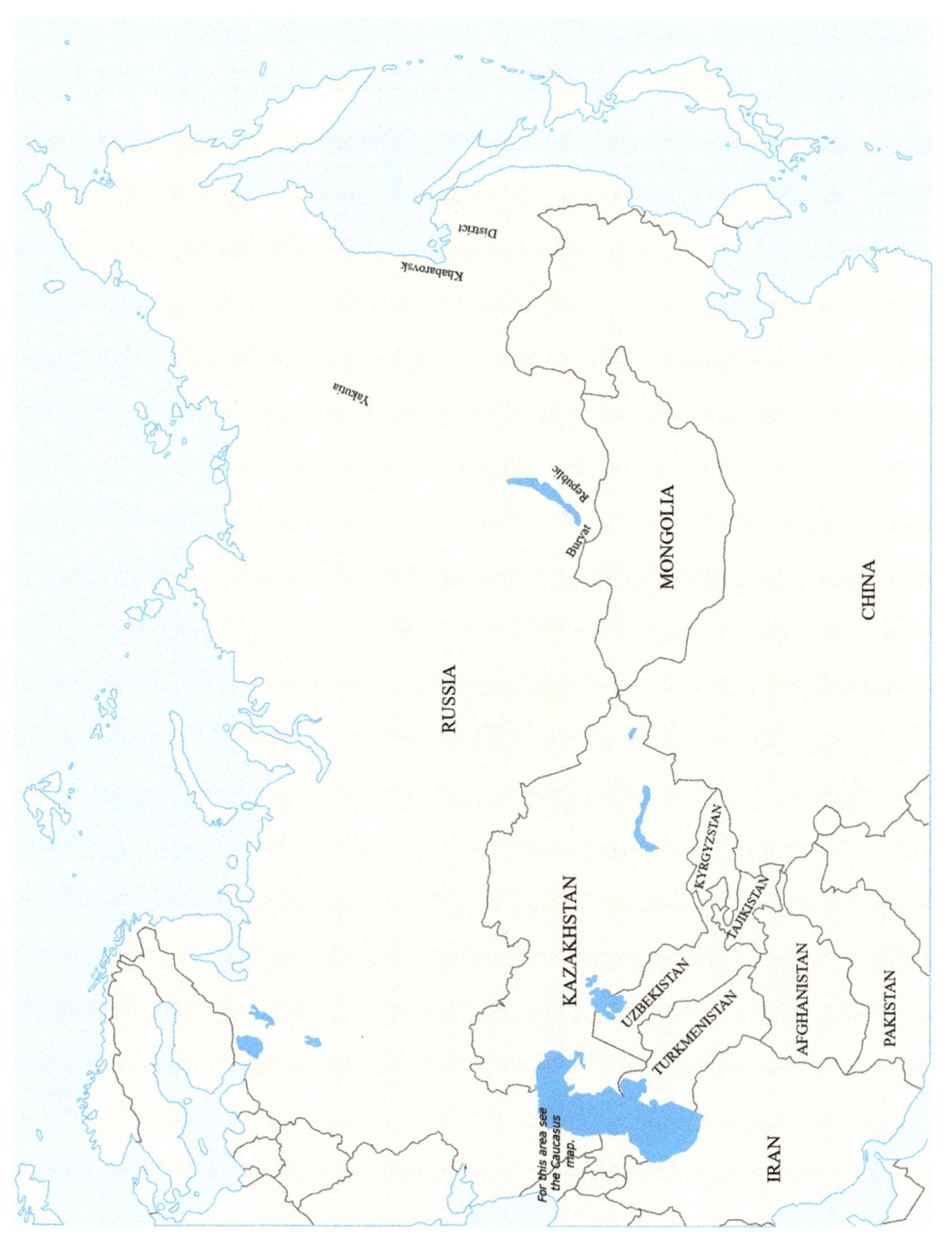

Central Asia

RUSSIA
KAZAKHSTAN
MONGOLIA
Dzungaria
KYRGYZSTAN
TIEN SHAN MTS
TAJIK-
ISTAN
Xinjiang
Kashgar
Tashkurgan
Gansu
Manchuria
BEIJING
N. KOREA
S. KOREA
Shaanxi
Tibet
NEPAL
BHUTAN
INDIA
Yunnan
MYANMAR
VIETNAM
LAOS

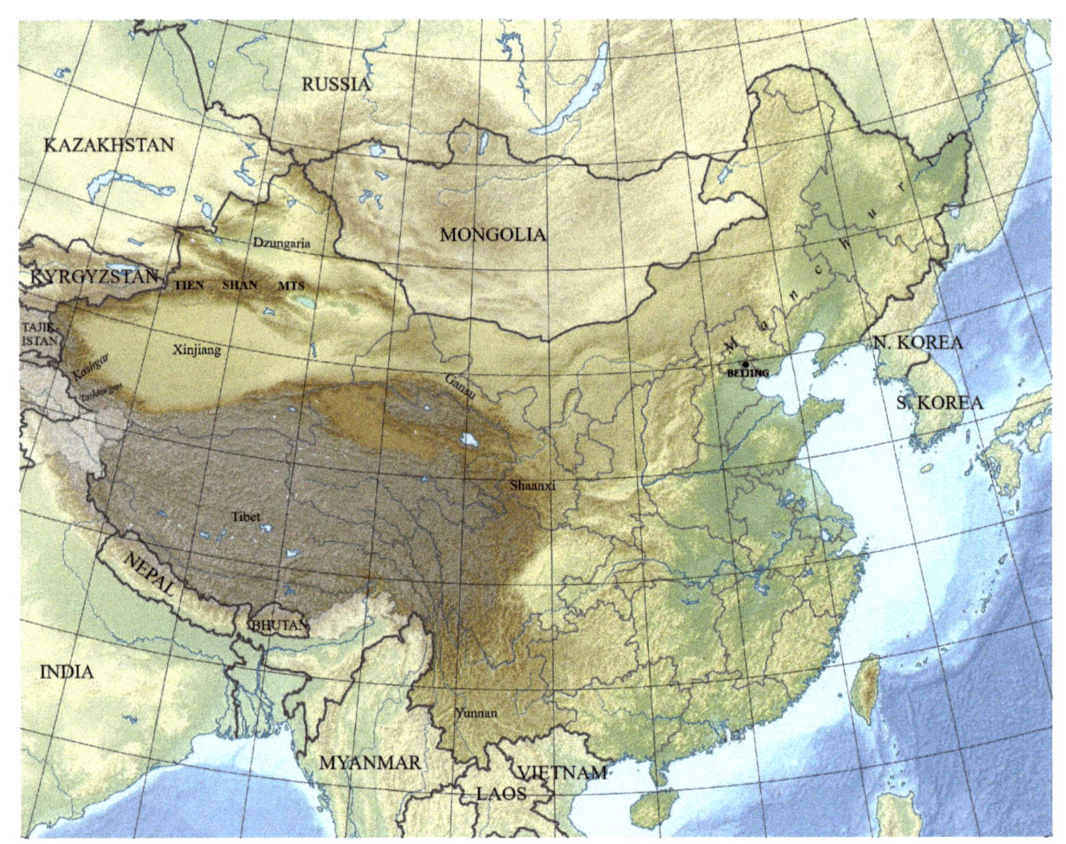

China

RUSSIA
Buryat Republic
W. Sayan Mts.
E. Sayan Mts.
ULAN BATOR
ALTAI MOUNTAINS
Gobi Altai Mountains
GOBI DESERT
CHINA

Mongolia

TIBET
LHASA
BHUTAN
Sikkim
NEPAL
KATHMANDU
INDIA
DELHI
Kashmir Region
Karakoram Range
Hindu Kush
ISLAMABAD
PAKISTAN
H I M A L A Y A S

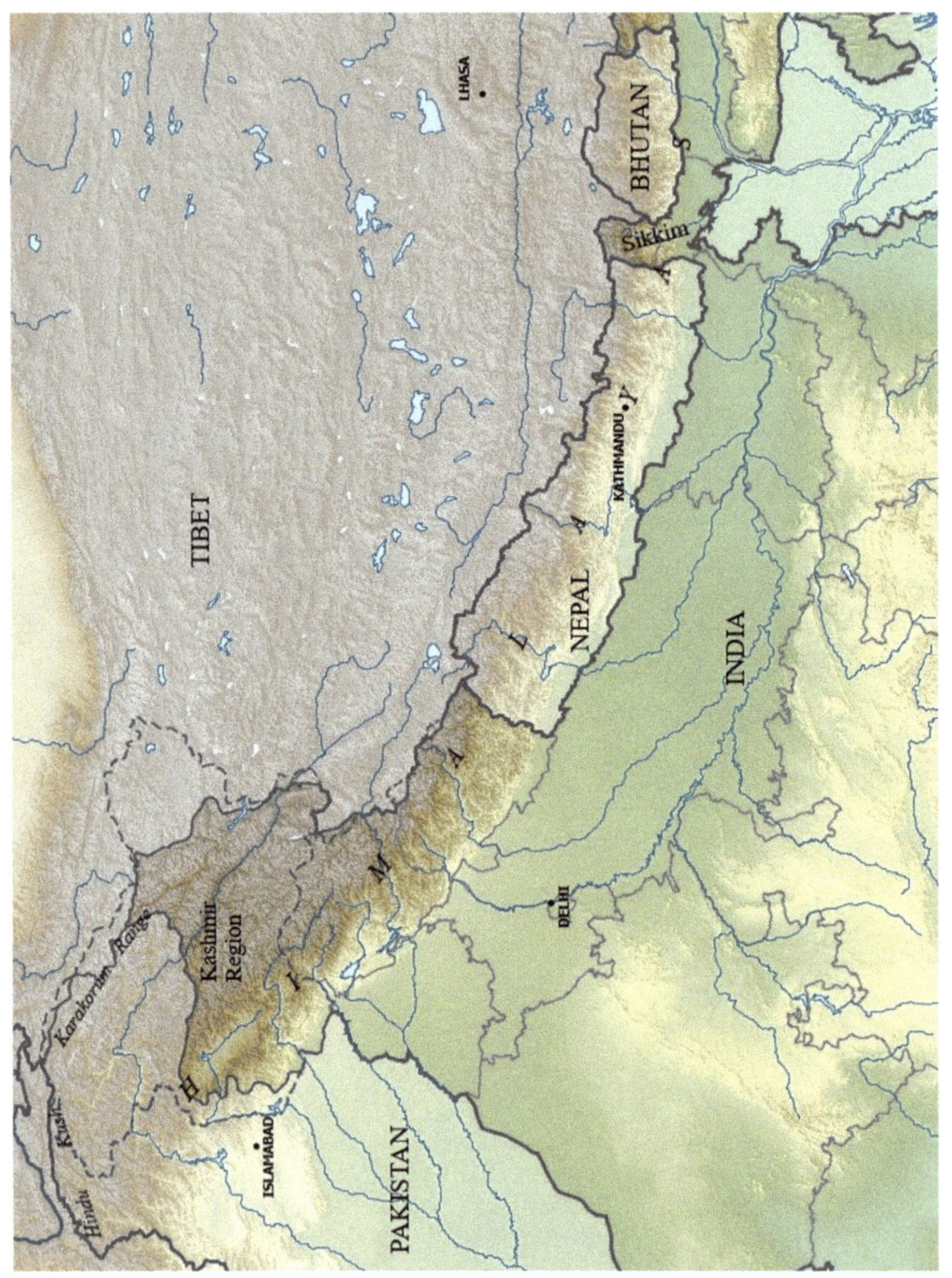

Himalayas

KAZAKHSTAN
TAJIKISTAN
UZBEKISTAN
CHINA
Chatkal Range
L.Sary-Chelek
Jalalabad
Arslanbob District
OSH
Kyrgyz
Alatau
Range
Suusamyr Range
Dzhungoltau
Range
Fergana Range
R.Kara-Balta
R.Aksu
BISHKEK
R.Narin
Lake Issyk-Kul
Engilcheck Glacier
Khan
Tengri
T I E N S H A N

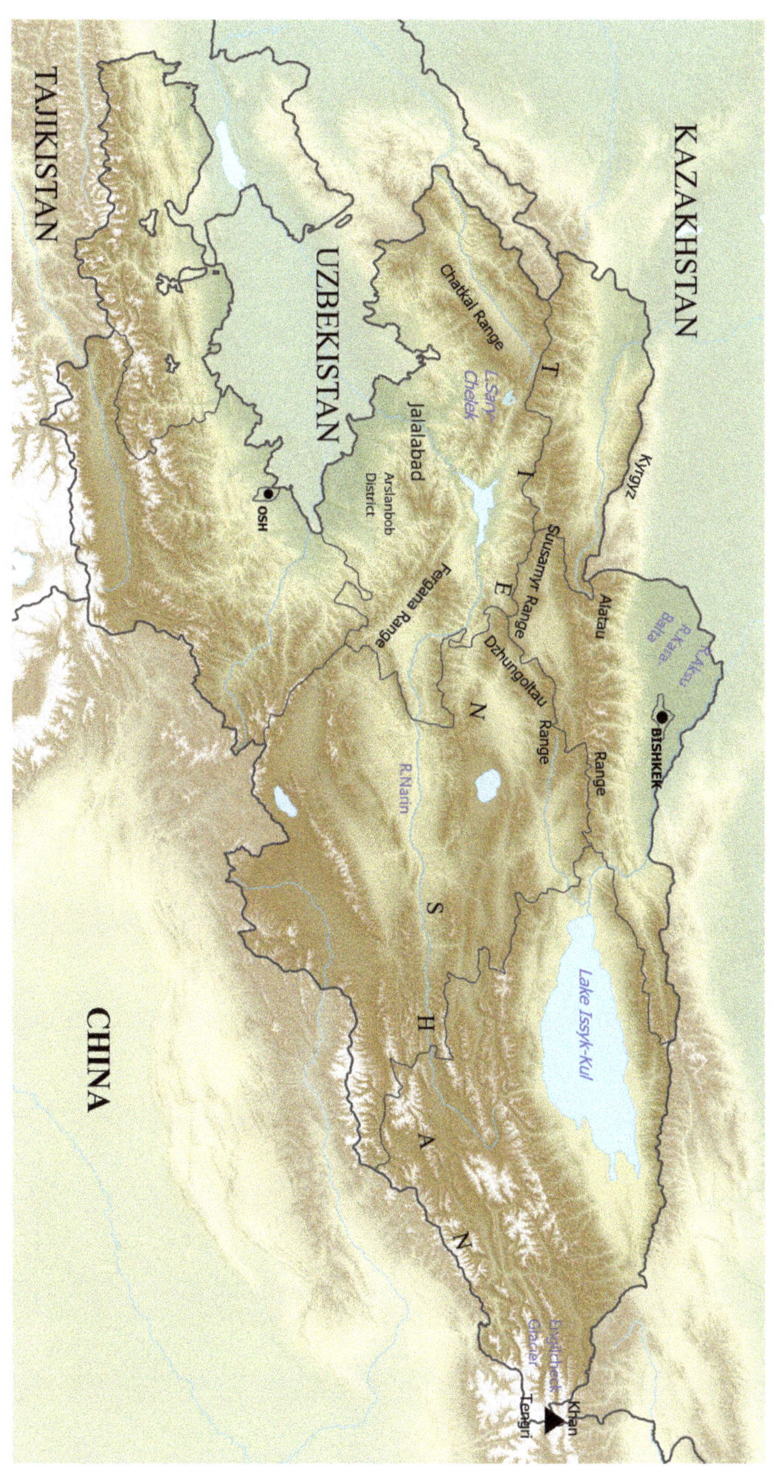

Kyrgyzstan

East Tajikistan - The Pamirs

West Tajikistan

Nepal & Sikkim

CHAPTER 9 • WESTERN ASIA AND THE CAUCASUS

Afghanistan

The amount of reports and material coming from this area of Asia is fairly small, but as there are some records from the area it is clear that even here there is, or has been, a species of higher bipedal primate.

The Soviet engineer M.N. Fokin, who was working in Afghanistan in 1953-1957, wrote the following account of a story told by one of his translators. In 1944 some Afghan border guards met a male and female ape in the bushes and reeds near the confluence of the Panj and Vanj Rivers. They shot the female, but the male hid itself in the reeds. The dead female was about 2 metres tall and covered with red hair. They were supposed to transfer the corpse to a museum, but later on it was purchased by the British. They did not manage to protect it against decay, so the fate of it is unknown.

Another curious message comes from an Afghan teacher who now lives in Turkmenistan in the Soviet Union. His name is Juma Alimadat. He was born and spent his childhood in the Duzahdara area in Afghanistan. As a child Alimadat heard many stories about the *gul-biyavan*, the wild man, who was covered with hair, and he was in no doubt about its real existence. "There are wild men in the wooded areas of Afghanistan where I was born. You met them alone and with young, and sometime in flocks. They live in the woods, and they feed on wild animals they kill with stones. Gul-biyavan does not speak. It can only mumble incomprehensibly." This report is very interesting, as it mentions young, as well as flocks or herds of these animals. This is something we have only heard once before, in a report from south-west Xinjiang, i.e. in a place not that far from the areas immediately east of the Hindu Kush.

Another report from areas adjacent to Afghanistan tells us again about observations of "wild men", not just alone or in pairs, but sometimes even in flocks. This report comes from F.F. Schultz, who in 1949 in Tashko had the opportunity to talk to a Tajik -Afghan agronomist and a shepherd from the border areas of the Pamirs, Afghanistan and India. They told him that people in the border villages would see hairy naked people searching for food. Sometimes people would set out fruit, cakes, meat or milk in bowls for them. They would usually appear in groups of 6, 8 or 12 individuals. The wild people only show themselves once every few years. It is a local custom not to drive them away without giving them food. According to the shepherd, they still resemble man, and as such are a creation of Allah, so to hurt or kill a wild man would be a great sin.

One day, a Soviet guard managed to get close to one of these naked wild men and throw a net over him. The other wild people were scared away when he fired shots into the air. However, he did not keep the wild man, but brought it back and released it at the same spot where it was caught, much to the satisfaction of the Muslim population.

The wild naked people who appear on the borders of the Soviet Union and Afghanistan "are the same size as us, but they are hunched over, and they use their hands a lot, especially when they climb in the mountains. They look like monkeys when they climb. Their hair is brown with hints of red and black. They have very little hair in the face. Here the skin is swarthy, almost black. The head has a sloping forehead, prominent jaw, and high cheekbones. They are quiet by nature, not ferocious. They have never attacked people. They collect their food in a peaceful way."

The following story comes from Kularova, a 70-year old Turkmen hunter living in the southern part of the Turkmen Soviet Socialist Republic, about a hundred kilometres from the Afghan border. "It was in Tsarist times. I went with a caravan to Afghanistan. It was winter and everything was covered with a thin layer of snow, but in the mountains the snow was deep. We had stopped and made a fire to cook tea. Suddenly we noticed some hairy beast coming towards our fire. Thinking it was a bear I grabbed a gun, and together with another hunter, we crawled towards it. When we got closer I saw it was not a bear, but a wild man – *javan-adam* or *demir dernmark* (man with iron claws). The wild man saw me, and started to move away from the fire on all fours. Wild men can run faster than a horse on all fours."

Unfortunately, we do not have information from all of Afghanistan, possibly because scientists in Afghanistan and Iran have not been interested in the problem. It may also be affected by the fact that we come close to the heart of the Muslim world when we move towards the west.

Iran

From the Hindu Kush, we get to the Central Afghanistan mountain watershed and through its northern branch, the Safid Range, to the ranges of Kopet Dag and Alborz. According to some of our preliminary data a significant number of observations of wild bipedal humanoids were made in the Soviet part of the Kopet Dag Mountains in the 1920s. With respect to the Iranian Kopet Dag as well as the Alborz mountain range, and the entire Iranian plateau, we have absolutely no recent specific information. Or rather, there are a couple of references dating back to the Second World War. Sanderson cites a letter from an Englishman who worked in Iran, saying that some Iranian had brought an animal resembling a gorilla from the mountains where they had killed it on a hunt. A.Z. Rosenfeld told the author of these lines that while working in Iran he also heard something similar.

If we look in ancient texts, we see indications pointing to the fact that Iran and maybe

Afghanistan was in fact the ancient centre of habitat for the humanoids, and that they spread to the adjacent mountain systems from there. In the 17th century the German scholar Athanasius Kircher wrote about humanoid bipedal hairy animals from the mountains of Southern China. Kircher did not give a detailed description but simply wrote that it was an animal like "Persian man (*Homo persus*)". A few hundred years ago Bigfoot was in other words not known to Europeans as an inhabitant of the Himalayas or Tibet, but as an inhabitant of Persia!

Another strange thing is the name *gul-biyavan*. We find it in two widely separated areas: in the Pamirs and in the Caucasus. Clearly the name must have come from a common root – from Iran, and therefore the creature must have been known for a longer period of time there.

There is no biogeographical objection to the possibility of pockets of endangered species still living in the Caucasus – even wild men. It is well known there has been a diffusion of species between the fauna of Central Asia and the Caucasus. The mountain ranges and highlands of northern Iran are a natural biogeographical bridge that would be perfectly capable of binding the Asian Bigfoot area together with the Caucasus. The area, called Mazandaran, is extremely mountainous, and in lower altitudes it is covered with subtropical forests. The average altitude is 3000 m. Many passes are around 2000 m, and the highest mountain, Damavand, is over 5000 m. The geological and climatic conditions have contributed to the small human population and relative isolation of Mazandaran, particularly the Alborz mountains, from the rest of Iran.

Count Nikolai Ants-Kars wrote in his "Memoirs" about a curious story he had heard in 1828 in the Caucasus from the former high chief of the Turkish-Persian army. Apparently an Iranian from Mazandaran had told the story of how as a child he was caught in a ravine and kidnapped by "divas or monsters", like the people remaining after the flood, with whom the current population had gone to war several times. Their main residence is to be found on Mt. Damavand in Mazandaran, the mountain won by King Solomon. They are mentioned in all the fairy tales with such detail that some European writers have even begun to assume that such an antediluvian breed of people could somehow be preserved. According to legend, in ancient times the war against them was led by the Persian king. As of now it is a complete mystery to us what "European writers" the count was talking about.

The legends mentioned here are open to a lot of research. However, these distorted and fantastic reflections on relict hominids in Eastern literature and folk tales are not part of the theme of this book. In the future, we will probably see philologists and orientalists engage themselves closely in this subject. A digression into the traditions of pre-ancient times may be worthwhile, as the myths are interwoven with real history, refracted through the prism of religious and mystical thinking.

In Chapter 2 we mentioned that the oldest written information about creatures that can be compared with Bigfoot we find, on the one hand, in the Assyrian-Babylonian Epic of Gilgamesh and the ancient Jewish texts of the Old Testament, and on the other hand in the Vedas and the Zend-Avesta. [48] This raises the question of whether these creatures entered Iran from the Middle East (or Arabia) or from India. In the poem of Gilgamesh they are known as *ea-bath*, *yaban* or *javan*, and in the Zend-Avesta as *divs* (*divas*). We are very much interested in beings that are known under those names to this day, among for instance the Iranian-speaking Tajiks. Research by A.Z. Rosenfeld has shown that the term *ghoul* (*hum*) is Arabic. If this is associated with the spread of Islam, it is of course much more recent than *ea-bath* and *div*. So, from where did the relict hominid move into Iran? From the East or the West? The answer to this question would be an important argument for a solution to the dispute about the nature of the biological and evolutionary place of Bigfoot. It is of course impossible to solve this problem by using ancient literature, but it is a question we must not neglect.

The ancient text of the Zend-Avesta was fully formed in the middle of the 2nd Millennium BC, but it probably started even earlier, i.e. it is probably close to 5-6000 years old. But the text itself says it was written 4000 years before the legendary king Yima Hshaeti. According to the Zend-Avesta, the ancient inhabitants of India waged a bitter war against the *divas*. They are huge, naked and hairy. They fight without weapons, using stones and trees they have pulled from the ground. The fight went on for three generations, but in the end the divas were partly exterminated or driven into the Alborz Mountains in Damavand. In the Pahlavi variant of the Avesta, which came to be under the Achaemenids, the divas continued from time to time to harm the Indo-Iranians even after the exploits of King Dzenshida had brought peace and prosperity to the people. Later on there was the story of King Zohak and the Assyrian-Babylonian invasion of ancient Iran. This epic conqueror had a connection with the divas. They had even taught him their cannibalistic ways, eating the human brain. He also had divas as part of his troops. In the Avesta, there is almost nothing supernatural about these creatures: they are of flesh and blood, the same as men, but their bodies are covered with hair, their have fearful faces, are bestial, know nothing about weapons, they fight hard and are dangerous, not because of magical strength but because of physical strength. In the Pahlavi version of the Avesta they are endowed with a few more fantastic features than in India, and in the Muslim folklore they finally turned into evil spirits, although mortal. Let us dwell on one very curious trait attributed to the divas in the Iranian epic: they are used in the army as a kind of trained fighting animals. The poet Nizami compares such use of the diva with the role of war elephants. Another Persian poet, Ferdowsi, tells how king Zohak ordered the rulers of the provinces to add divas and *peri* (winged spirits) to the troops. *

* *Translator's note: In this part of the manuscript there is a page and a half of notes about the so called "divas" in ancient literature. The notes are very short and disjointed, and one must presume the author had intended to expand this part on a later date, but never got around to do it. I have therefore decided not to include them in this translation.*

Talysh

We have now come to one of the most difficult parts of our geographical survey: the Caucasus Mountains and data on whether there is a population of relic bipedal humanoid beings – in general similar to the yeti – living in the Talysh Mountains and in various places on the main Caucasian Range.

Hardly any other page of the Bigfoot epic has caused so much controversy, protest and ridicule as the suggestion that the creature can be found in the Caucasus. The entire Caucasus area is well described in zoological terms! It is well known and people have walked all over it! This view is however very far from the truth. There are many almost inaccessible areas in the Caucasus, and there are many unsolved zoological problems, such as the existence of pikas, ermine, the spread of the leopard etc. There is still the possibility of serious discoveries to be made.

However, if we deal with only folklore and traditions, are there stories of possibly long extinct great apelike beings here? Maybe traces of the real features of this animal have merged with legends, and it may be possible to pick out some grain of reality from the great mass of mythology.

As a start, we can search the published literature. Professor A.A. Mashkovtsev, a prominent researcher of the issue, has written: “The processing of the collected literature and material from our questionnaires from the Caucasus has exposed some very interesting data”. At the end of the last century, from 1870 to 1899 and again from 1915 to 1936, about twenty papers and two books were published in which the authors give very realistic descriptions of “wild and hairy forest people”, who, according to both the people of the Northern Caucasus (Chechens, Ingush, Karbadin, Karachay Tartars etc.) and the people of the Black Sea Coast of the Caucasus (Circassians, Abkhazians, Mingrelians, Georgians etc.) still live in the dense impenetrable forests of the Caucasus.

In the middle of the 19th Century, the head of a British expedition trying to find the victims of... wrote a report to the Royal Geographical Society, in which he gave information from Russian officials about wildpeople living in the woods. [49] Tradition says that in the old days, when the Abkhazians had just appeared in the country, the *abnauayu* or *ocho kochi* (forest people) could be found throughout the forests of Abkhazia and newcomers had to endure a bitter struggle with them. Imeretians described *ocho kochi* as a biped entirely covered in hair. Most of these stories were written and published by ethnographers, and accordingly they are refracted through the prism of ethnographic thinking. We will return to this later, but first we must turn to the work of the author and zoologist K.A. Satunin, who was a famous explorer and researcher, specialising in the mammalian fauna of the Caucasus, and the so-called *biaban-guls*. Satunin reported on his own field observations, but also on data received from residents.

Late on a night in March 1899 in the southern part of the Lankaran district in the swampy foothills of the Talysh Mountains, Satunin was on his way on horseback with several guides through a forested area in an almost impassable marshy swamp. Suddenly the horse stopped and acted very nervously, with twitching ears. Satunin peered forward and clearly saw a dark figure in the twilight crossing the road and silently disappearing in the thicket. It had "completely human movements". One should recall that Satunin's numerous works have established that he is a first-class observer. His chief guide Kakhiani reached for his gun, but threw it behind his back, when the figure disappeared. "We were in such a sober mood, that it is difficult to see why we should have experienced a hallucination. Besides, the horses undoubtedly also saw something."

It was only when they had finished their journey and were resting in the house of the village elders, that Satunin's guides told him about the "wild people" living in the Talysh Mountains. According to one of the guides, the wild men were already extinct, and all that were left were the wild women, and it had been one of them they had met earlier on. Another guide claimed that the wild men could still be found in the area. One day, after having grazed the horses in the forest, he had gone hunting, and had met a very tall man. His body, head and hands were like a man, but his whole body was covered with hair, and he had long fingernails. When approaching the hunter, the wild man had made strange convulsive movements ("as if he wanted to scare me"), then screamed and fled into the woods. On another occasion, the same wild man had grabbed the hunter's neck and leg, but had run away when he heard the cry of approaching people. The third time the guide met with the wild man, he fired his gun at a tall shaggy man, who was walking on the trail in the forest. According to the story he hit the wild man, who shouted, waved his hands, fell to the ground struggling, jumped up, fell down again, jumped up yet another time, and then fled. In the morning five villagers and the guide found a lot of blood on the ground. They trailed the wild man for three days, but in the end had to abandon the hunt.

Satunin has also recorded another story told to him by an old collective farmer called Allahverdi Mirzaev. "The old man told me that in the mountains (Talysh Mountains in Azerbaijan) people sometimes met a humanoid creature much larger than a man, and covered with hair. The locals called it *guleybany*. The old man told me that he had heard a thunderous roar in the mountains at night, a little like a human cry… One hunter who had met the creature in the forest at first did not shoot him for religious reasons, but later decided to shoot the guleybany anyway, and did so at close range. It made a thunderous cry and fled into the forest. The hunter went back to the village and told about shooting the guleybany. A group of them went to the place and found blood. When they followed the trail, they found a dead very tall creature like a human being, with long arms and thickly overgrown with hair." There are also modern stories told by residents in the same area about being attacked by guleybany hiding in the swamps.

Captain Belalov, who is employed in the Balakan District Police Department in the Azerbaijan Soviet Socialist Republic, has reported on a case from the summer of 1947, about a police sergeant who served with him in the police department in the Astara region of Azerbaijan SSR near the Iranian border. "One night, when returning from duty, the sergeant had been the victim of a strange attack, and went home sick, frightened and scratched. After 10 days he recovered, came back to work, and told us the following: He was returning home late one night, when the moon was shining brightly. He was armed with a gun, and was walking calmly towards his village not far from Astara. In front of the village there is a small river, which you cross on a bridge. A few trees are growing near the bridge. As soon as the sergeant passed the bridge, a huge hairy creature jumped out from a tree, and started hugging the sergeant with its hands. There was a second creature near the tree. Both of them made strange inarticulate sounds. They seemed very curious, and started to look and feel at the man, especially the shiny buttons on his uniform and his face. The sergeant was so scared he forgot about his weapon, but he did not lose consciousness. He said that what he saw in front of him was two huge wild people covered with dense dark hair. They were a man and a woman. The woman was a little smaller than the man. She had long hanging breasts, and long hair on the head. Neither of the two had hair on the face, but their faces were very scary. The skin was dark, and they looked like the faces of monkeys. The sergeant assured us that he could see all these details because the moon was shining very bright. As soon as the sergeant started to move, the male growled menacingly and pounced on him, and scratched his face. The sergeant was very scared, but he remembered the end of the terrible night in detail. When the female started touching the sergeant, the male started growling at the female and tried to drag her away. Maybe he was showing some form of jealousy. When the male pounced on the sergeant, the female started to growl and tried to push him up from the ground. At one time, the two wild people even fought among themselves. As soon as dawn broke, the two humanoids disappeared. And after spending a few more minutes on the spot, the sergeant ran home. When he arrived home he came to his senses and remembered his gun, and also that he had left his cap at the scene. In the morning he sent his children out to get his cap, and they brought it back. After spending 10 days at home recovering, the sergeant went back to work and told us this story."

From 1961 we have hundreds of stories about the population of wild men living in this area according to the local population. They call the males *guleybany* and females *vilmozhin*. Ancient customs supposedly prohibit local hunters from killing them. It is said that in autumn the creatures appear near the villages in melon fields and gardens. In the summer, they live near rivers teeming with fish, frogs and crabs. All the informants said that the creatures are larger than humans, that they walk on two legs, have thick brown or yellowish fur all over the body, very long arms, and large teeth. A group of Soviet zoologists discovered a fresh trail of a wild adult female with a baby. They made sketches of the tracks (print length 31 cm). They tried following the track, but because of approaching darkness had to stop.

North Azerbaijan
In the areas adjacent to the Greater Caucasus Mountain Range the hairy humanoid creatures usually have other names: *vehshi-Adam* or *mesh-Adam*. There are also other seemingly somewhat strange terms, *hyena* and *kaptar* (or *kaftar*) – this though can be found in other areas, where it is used as a synonym for "wild animal".

In the summer of 1959, a number of people living in the mountains in the Balakan, Qakh and Zaqatala districts of the Azerbaijan SSR were interviewed regarding their understanding of the "wild man" of the main Caucasian Range. Most of these records have not been published, but we can present a short summary here of the data collected by M.-J. Kofman. First it is necessary to note the different attitudes of the local people regarding the stories. Many people do not believe in them, and regard them as fairy tales. Others, including some responsible local workers, think that these creatures represent "wild" people, i.e. people who have left human society in distant or recent times, such as during the Caucasus wars of the 19th century, during the Revolution, or in subsequent years, including people hiding in the mountains to avoid tribal revenge or prosecutions. The Muslim clergy, the mullahs, attach some sort of mystical dimension to these human-like creatures, and are a strong hindrance to collection of information about them. It is quite interesting that Kofman remarks that the local population in general have a very poor knowledge of the surrounding wildlife. Hunters know only of those representatives of the fauna which have a hunting or fishing interest (bear, wild boar, jackal, pheasant and so on) or a commercial value.

From the more than forty records of encounters with the "wild man" in these parts of the Azerbaijan SSR, it is possible to formulate a tentative description of the creature. The fur is smooth and thick, red, brown or black in colour. The height is not less than 1.8 m, and in some cases more, up to 2 or even 2.1 m. There is no significant difference in build between the "wild man" and humans, although they are often described as having very long fingers, and in one case very prominent lower tusks. There is though a number of stories where the creatures are described as being smaller than humans, with long arms that reach the knees and a strong stoop. These animals have hardly any neck, so the head seems to sit directly on the shoulders. The head is also elongated and almost egg-shaped. Kofman was also able to show that, over the past five years, seven out of thirteen sightings have taken place in the mountains and forests in the Alazani River Valley. A series of sighting from 1947-1950 were distributed with four sightings in the spring, ten in summer and early autumn, and two in late autumn and winter. The same sightings took place at nightfall (seven), late evening and morning (three each) and three during the day. The last three had no specific time data. In nine out of the sixteen cases, it was the "forest man" who approached the eyewitness, including three cases where it came close to a fire. In the other seven cases, the eyewitness accidentally found the "forest man". There was no sign of aggression in any of the cases. Of the creatures sighted six were definitely male and two female. In the other cases the sex of the animals was not given.

The following are examples of these sightings.

- Navrus Guliyev, 35, saw a "man" covered with 5-6 cm long red hair late one night in 1952. He was standing on bent legs, but according to the eyewitness he would have been very tall had he been standing upright. He had a very broad chest, broad shoulders, powerful muscular arms that were longer than a man's arms. The hands were larger than a human hand and had a little hair on the back. The face was like a Kalmyk or a monkey. The teeth were large, at least twice the size of human teeth.

- Nurmamedov Rajab, a farmer, said that in April 1959, having gone out in the morning to collect firewood in the mountains, he saw about 40 metres up the side of the mountain a creature, which he at first took to be a bear until he saw that it was standing upright on two legs. Its head was covered in thick long hair. When he hailed it, it turned and went slowly into the forest on two legs.

- Gamsat Hisiev, a farmer, said that in the summer of 1955 or 1956 he had seen a humanoid creature in the forest at a distance of 25 metres. It had no clothes, and was covered with dark grey hair. It had very long hair hanging from its head to its chest. It was built like a man but seemed thin. The back sides of the hands were hairy. The "man" walked slowly into the forest, and the eyewitness followed him at a distance of 25 - 30 metres. When they came to a river, the "man" waded across, and then disappeared into the woods. The eyewitness did not follow.

- Kamal Aliyev, a bee-keeper and a hunter, said that in September 1957 he was spending a night in the mountains sleeping next to a campfire. He woke up suddenly in the middle of the night, and saw a human-like figure two metres away. It was about 1.8 m in height, well-developed, without clothes, male, entirely covered with thick chestnut-coloured hair. It was especially long on the head, but there was no hair on the face. The creature stood perfectly still with hanging arms, and showed no hostile intentions. When Aliyev fired a gun, the "wild man" calmly walked into the forest.

- Geysar Bakhshiyev, 34, a bricklayer, saw a creature resembling a man in March 1959 near a river. It was crawling on all fours, but then lay down flat in the water before rising and walking to the opposite bank. According to the description it was about 2 m in height, had thin long fingers and hands, and it was covered with brown hair. "His face was midway between man and ape."

- Nazhmatyk Gudule, 49, a driver, saw in 1955, at dawn in a forest, at a distance of about 15 metres, what he considered to be a man without clothes, but completely covered with thick brown hair. It was not a bear, as it was standing on two feet. It first rushed to one side of the road into some bushes, and then across to the other side, before disappearing into the woods, still running on two legs. The witness mainly saw the creature from the back, and but noticed especially the long hair on the head.

- Ekabashov Shirin, 22, a farmer, says that in October 1959 he was riding in a cart pulled by buffaloes when a tall hairy man appeared at the side of the road, 3-4 metres from the cart, and then ran across the road on two legs. The creature was covered in dark fur, 5 – 6 cm long.

- Izrafid Azimov, 47, a shepherd, says that in March 1959 in the Alazani Valley, a "man" came to bask near his campfire. His fur had a grey-brown colour. There was no hair on the palm of his hands, which he held out to the fire. "His hands were long. His face looked like a monkey. The hair was like a matted mane. The neck was short, and the head was sitting straight on the shoulder. He looked like a great ape, probably 1.8 m in height".

The questionnaire material is quite large, so we could give many examples. However, the ones given already give a good idea of the nature of the reports. Many of those interviewed described the cries of *mesh-Adam* in detail, and others gave specific information about their tracks. According to one description, their footprints looked very much like a bare human foot, apart from the fact that the big toe was placed off to one side. The length of one print is about 35 cm, usually with clear imprints of five toes and the heels. One usually also finds imprints of long nails.

Here are some more examples... *

Israfil Khalilov, 67, had been storing lime with Shamil Algaevyn Kazbina, 18, in the summer of 1945, in an area near the forest. "In the evening we made a fire, cooked meat. Then a man appeared, walked around the pile of stones close to us, and sat on a rock ten metres away from the fire and two metres away from Shamil, who was sitting at a distance. At first we thought it was a man who had come to scare us, and to this end had put his coat on inside out. Then we saw that it was not a coat but real fur, covering the whole body ... The man took a large stone in his right hand, and beat it on the stone on which he sat. He struck two or three times, and then the hard stone was crushed to pieces ... He did not say anything. The boy threw him a piece of bread, he took it and continued to sit. So we sat for a long time, an hour, maybe more. We were talking quietly, as he sat, looking around. When we looked at him and started talking to him, he did not react. Then we wanted to sleep, but did not know what to do. I said to the boy: 'Give me the gun.' As soon as I took the gun, he immediately fled - on two legs like a man, but as he ran on, he went down on all fours. He was

* *Translator's note: in this place there is an almost illegible sentence in the manuscript, possibly indicating that these reports could be included in the finished book.*

slightly higher than average human and thin. The coat was black, or so it seemed in the darkness. It was straight, not curly, and long. He had particularly long hair on the head. It covered the whole face, so we have not seen the forehead, eyebrows or ears. He looked human. The hands are very long, and very large. The nails are long, 4-5 cm. The toes we did not see."

In March 1959, Jamal Lativov, 32, went to the forest to collect firewood, 7-8 kilometres from the village of Kish. He arrived at his work area around 11-12. The forest was sparse, and he suddenly saw something similar to a man coming down from the mountain. It was completely covered in hair. It was going directly towards Lativov, but stopped around 10 metres away. They stood looking at each other for a long time, perhaps half an hour, when the hairy man suddenly turned, and walked calmly away in the direction he came from. "It had no clothes on. It was standing up straight on two legs like a human. It was of medium height, but with a very broad back, broad chest and powerful arms. The hands and fingers were large. Nails were long and wide, not compressed and bent like the claws of a bear. The shin and upper foot was covered with thick, long and seemingly very coarse hair. The face was hard to see because of the long, tangled hair. I could only see his eyes, nose and mouth. He looked at me like a bull. The eyes were small, but not very deep-set. Where I could see the skin, it was very black. The ears were black too. The teeth are smooth and white, no bigger than a man. The fur is very thick, but it is hard to tell how long, in my opinion less than 8 centimetres. It seemed to me it had more hair on the chest than on the back. The hairs were shorter on the buttocks, and the skin could be seen. On the lower back the hair was very thick and long. The neck was not visible at all. Judging by the lack of breasts and a powerful physique, I think it was a man, but the stomach and abdomen were covered with thick hair."

But Lativov's adventure was far from over. In spite of the fear, he decided to start working in order not to lose a whole day. But he had hardly cut down the first tree before the creature appeared again, and this time coming very fast towards him. Lativov feared it would attack him, so he ran away and climbed up into a tree with an axe in his belt. "The hairy man came up to the tree and stood there, probably for about two hours. My hands got numb from holding on to the branches. From time to time he would look at me, and then look around. He never made a single sound. Sometimes he would fold his hands, and lay them on his head. I had the impression that the first time he was calm, but the second time he wanted something from me. Then he turned and left, this time along the river. I had never heard of such a creature, and I do not know what it is, and what it is called. Were it not for the fur and the long nails, it could almost be a man."

Naurudov Ramadan, 48, farmer from the Azerbaijani village, Poshtbina, said that in 1951, when working as a watchman in a warehouse of nuts, he once, early in the morning in September, heard splashing in the river and decided to go check. "I saw a woman standing in the river, her whole body was covered with black fur. The hair on the head was very long. Her breasts were hanging down to the lower abdomen." As he was armed with a small calibre rifle, Naurudov decided to kill the hairy woman, but when a branch broke under his foot, the hairy woman fled very quickly into the bush, running on the legs like a human.

Mustafa Mustafayev, 50, says that in September 1954, in the area of Jean-Jar-Tala, he was riding a horse on the edge of the forest when he saw a creature resembling a human. "It walked on two legs, slowly, across the field toward the woods, to intercept me at a distance of approximately 70 - 80 metres. It was taller than a man, and covered with hair like a bear. A few days later Mustafayev saw the same creature again. It was walking on two legs very slowly. It was covered with red hair. In December of the same year, Mustafayev saw the same "forest man". "He shouted very loudly two or three times." The creature has been seen in the area for at least 3 or 4 years. Mustafayev also claims that he could easily have caught a forest man, but the idea had never occurred to him, as he thought science knew all about the creature.

Khalil Khalilov, 35, an Azerbaijani forester, is known in his village of Bash-Kyungot as a very skilled and brave hunter. In May 1954, he was coming home about two in the morning with some other villagers. The night was quite dark, and the travellers only noticed a man walking slowly forward when he was 25 metres away. When he passed their donkey, he suddenly rushed to the side of the road and disappeared. Khalilov walked towards the edge of the road as well, and saw the creature sitting there. "It was hunched and covering its chest with its hands. It was leaning on a pile of sand to be used in repairs of the road." It did not react to questions or shouts, so the witness started throwing stones at it. Every time it was hit, it just moaned softly. Then it slowly crawled away through the ditch on all fours and under a bush and away. The witness did not see any details, but again expressed wonder that science did not know of this creature, and said he could easily have caught it.

[Translator's note: In this part of the book, there are several very short sightings of a long-haired, red or black hairy creature in the same area. The sightings do not have many details, and it seems the author intended to get back to them, and write them in a more detailed form later. I have decided to omit them, as they do not contain much information.]

There are several sightings of the creature *mesh-Adam* bathing in the river, throwing handfuls of water on its face or on its fur, often for a very long time.

Another story comes from the same area – the village of Poshtbina . Collective farmer Osman Osman, 32, said that in May 1947 he was out looking for his buffalo, and was about to cross the river on a thick fallen tree trunk, when he saw a man-like hairy being sitting on the middle of the trunk. It was covered with long black and red hair, similar to the hair of a buffalo. He was sitting on his haunches with his hands on his knees. The fingers on his hands and feet were long. The nails were long as well. It had the face of a man, "but at the same time looked like a monkey". It was thin, and not very tall, perhaps the size of an eight year old child. The eyewitness was standing for a few moments about 2 metres from the creature he called a *kaptar*, but then it suddenly jumped into the water with incredible speed and disappeared. "I myself lost my balance and fell into the river. It was the month of May, and the water was very cold. I caught a bad cold and was sick for a month because of this kaptar".

One of the more legendary stories about the kaptar is this one, where it is said that it can ride a horse. We do not have exact time or place for this story though. It happened somewhere on the Poshtbina collective farm. Waal Aliyev, 60, said that one day many years ago ("before collectivisation"), when he was still working for a master, he was ordered to capture a kaptar. It had been bothering a horse for many nights and chasing it around, and it should be taught a lesson. So Aliyev smeared tar on a tarpaulin, attached it firmly to the horse's back and let it graze freely at night. The first two nights nothing happened, but the third morning there was a female kaptar sitting on the back of the horse, stuck to the tar. It was very similar to a human, but the whole body was covered with red-auburn hair and its breasts reached to the lower part of the abdomen. The fingers and nails were long. It had teeth like a person, and the ears were elongated a little bit upwards. The people tore off the tarp and kept the kaptar for 2-3 hours, and then released it. After that, the horse was left in peace.

As a kind of conclusion to this group of sightings we might give the statement of the foreman of the Lenin collective farm in the village of Calma, Debir Musayev, 50. "Why has it taken so long for you to come and look for the kaptar? Why have your scientists not done so before? There used to be a lot of them. They lived in the forest three kilometres from here in a place called Nadzhindzhar Tal. When I was 16 years old, we often heard their cries on our farm. I was frightened of their loud cries, but my mother said it was nothing, only the cries of the mesh-Adam. The screams were very sharp and strong. They could be heard for 500 metres. They sounded a little bit like a human voice. Our people did not pay attention to them – we let them live, because they were not harmful. They were often seen high up in the trees. One old man, Hizriev Mamedov, who lived a distance from us in a house completely surrounded by forest, often complained to me because they would go very near his house. But then everybody started hunting, and the *mesh-Adams* withdrew. Perhaps they all died or moved away to the mountains. Mesh-Adam is a very clever animal. He is much smarter than a wolf. He rarely shows himself to man, and he can run very quickly." (Recorded 7 September 1960).

It may be possible to search for their wintering dens in the forests of the floodplain.

In a fairly limited area of the Balakan District, Marie-Jeanne Kofman and S.M. Mereshinsky collected a fair bit of information about a very special variety of hairy humanoid creature: a white kaptar. Apart from the unusual colouring, the white kaptar was also thinner, had more slender limbs, and a passion for ponds and small rivers. It would also come near human habitations and visit gardens, orchards, cornfields and grazing areas with horses. Witnesses attributed a number of strange properties to these creatures, which suggests that we are dealing with local folk mythological images. Here are two of the reports:

Mamed Omarovich Alibayov, 30, a carpenter from the village of Khullar, was considered the best hunter in the district, a brave and honest man. He met a kaptar in September 1956 about two in the morning, when he returned from a hunting trip, coming down from the mountains towards Khullar. About one kilometre from the village, he saw what he thought was a man sitting huddled on the opposite bank of the river. Peering closely, the witness discovered that the man was completely white. The hunter shouted, and the man started to move, but he did not get up, so the hunter started moving towards the stone the white man was sitting on. After a few moments, the white man suddenly stood up, stretched out his right hand, and putting it against a rock in the water, leaned heavily towards the hunter and made a quick and soft hissing or whistling sound like “shhh-shh”. He was the height of a normal man. The hunter jumped up and ran home, woke up his brother Alibek and came back with him and a lantern. The stone where the kaptar had been sitting was still wet, but he was gone, probably upstream.

Alibayov stressed that he saw mainly the face and right arm of the kaptar. "It was the face of a human, and not covered with hair, but it was almost completely hidden by thick grey-white hair hanging down from the head. When the creature leaned forward, his lips were moving very quickly. The hand was long and thin. When he got up, he seemed long, thin and stooping. The legs were very thin. Height perhaps 1.60 – 1.70 m. The whole of the body was covered with white hair. "I always thought that the state knew about the kaptar. After all, people talked openly about it. How could I possibly believe that the state and the scientists did not know. I was sure everyone knew about the kaptar just as they know about the bear, the boar and other animals. I thought it had been in zoos and museums for a long time."

Mogmatili Nurmagama Yunkurova, 24, a native of the same village of Khullar (now working in a tyre plant in Leningrad). As a child, he heard a lot about kaptar living in the surrounding forests and mountains. They were similar to monkeys, but were hairy wild men covered with thick white hair. His grandmother Karimat Mustafa Yunkurova Kesey told him that she had seen such a creature several times near the neighbouring village Poshtbina. In the summer of 1958 a kaptar had been chasing a mare in a very large garden. In the morning the horse was tired and sweaty, and its mane would be braided, although in a very ugly and messy way. The witness tried to tie the horse to a long chain, but the next morning the horse was again sweaty and hungry, the chain had disappeared, and the animal was tied to a rope. The chain was found a few days later at the other end of the site. On the 12 August 1959 the witness tied the horse near to the house, and went to sleep on the porch. About two he woke up, hearing the barking of a dog. When he stood up, he could see a kaptar sitting on the fence. The creature was like a man, but the hands, feet and chest were covered with thick white hair, and the hair on the head was also white and very long.

In the summer of 1953, he was also on the way to the village of Poshtbina with a friend, when they saw, at a distance of about 40 – 50 metres, two kaptar swimming in the river. Both were covered with white fur. They seemed to be washing or grooming each other. The next day some more people went to the same place, and saw the two kaptar again.

Asadullah Debirova, student, 12, was near the river close to Poshtbina in August 1959 with a friend. They heard strange sounds coming from the river, so they walked closer. When they parted the bushes, they saw a man standing in the water. He was all covered with white fur, but with long greyish hair on his head. He was the same size as a normal man, but had very thin legs.

Assad Shirinov, 17, from the village of Chamboulive, was out early one morning in January 1960 with a friend when they saw a kaptar approaching them on the road about 15 metres away. He was walking on two legs, was all white, and had no clothes. He was taller than a normal man, and had long hands with long fingers. When the boys shouted at him, he turned and walked slowly away.

Nazmiat Mamedow, 32, from the village of Magalamar, says that in October 1953 she was riding with the driver in the cab of a truck, and her husband was in the back. It was late at night. "Near the Katekhchay River, a kaptar suddenly appeared on the road. He was the size of a tall man, and covered with long white hair. The hairs on the chest and shoulders were long, but were shorter on the arms. He was jumping up and down and waving his arms. The driver got scared and stopped the car. We could see the kaptar closely, because he was standing in the light, and he was only 15 metres from us. It was a man. He stood there for a moment, then ran into the bushes on the left, and shouted very loudly."

It is probably natural to assume that we are dealing with a local group of albino mutants in this area. Perhaps only 5-6 individuals. Their albinism may be accompanied by other morphological and biological anomalies, but we cannot be sure.

In the summer of 1960 and 1961, E.K. Iordanishivili, D.I. Sergeeva, I.A. Vasilyea and E.G. Plachenova surveyed several routes from a zoological and tourist point of view. Among others they carefully examined some remote areas of the Zaqatala Reserve in the Caucasus Mountains. In 1960, near some deer trails, they found two prints that for a number of reasons can be attributed to the kaptar. The tracks were repeatedly photographed. There was a flat and massive first finger (toe?), and various other features that would attract attention, when comparing the tracks to those of the Bigfoot. The tracks were not similar to bear tracks in any way. In July of the same year, they were in the area of Kamsa Chai and Chai Dzhahazu, where they found mature bushes torn from the ground, and big branches torn from trees. All of this was photographed thoroughly. In the summer of 1961, in a different area of the Zaqatala Reserve, the group found very deep humanoid footprints on the shore of a river where a creature had jumped down from a stone. These were photographed, and show many similarities to tracks taken in California by Ivan T. Sanderson. In three places the group collected stool samples similar to human in form, but they were found in an area virtually without humans. These are only preliminary data, but will serve as preparation for future field studies.

Georgia

In the Lagodekhi district of the Georgian SSR, we have collected a number of data on the "wild" or "forest" man. I will give some examples of them shortly, but it is important to note that researching the Caucasus area is more difficult than anywhere else because of the many local names of the creature. In some locations people use two or three different names. Thus we have the Azerbaijani names of mesh-Adam or kaptar as mentioned above, but there is also *tyuhli-Adam* (hairy man), *kara-Adam* (black man) and other terms. In the surrounding area of Eastern Georgia people are more likely to talk about *tkis-katsi* (forest people) while in Imereti, Guria or Samegrelo it is *ocho pintre*, *ocho kochi* or others. In Dagestan they use names like kaptar, *rukh-chuzha* and many others. Collecting these names does not enrich the biological sciences in any way, but knowledge of them is a practical necessity for field work.

There is a fairly limited Muslim population in the area, so one of the general obstacles towards information gathering or obtaining a specimen is not a problem here. Although one does hear stories about hunters having shot specimens, when said hunter is interviewed they claim never to have seen such a creature or even heard of it. Perhaps fear of legal liability?

Gabri M. Apakidze, Georgian, 85, village of Kavshiri, and Nikolai Yasunovich Kapanadze, Georgian, 47, village of Thunder. Both told me that about five years ago, probably some time in September, they were staying in a hut in the woods collecting timber. Their hut was located 15 metres from a spring, which was the only source of water in a large area in the mountains. Early one morning Kapanadze, who was sitting on the doorstep, saw a humanoid beast coming to the spring, where it dropped to its hands and knees, and started to drink. It drank for at very long time, about 15 minutes, occasionally lifting its head and looking around. The creature clearly saw the hut, the men and their fire, but kept drinking. Finally it left, looked over its shoulder at the hut, and started going back up the slope where it came from. Kapanadze had awakened the older man, and asked what kind of creature it was, as it was not a bear, but the old man whispered: "Be quiet, be quiet, I know what it is."

The stranger looked like a man walking on two legs, but he had no clothes on. From head to toe he was covered in long dark, reddish hair. He was taller than a man, but a little stooped. The bent arms looked longer than human arms. The fingers and nails were very long. He walked with a normal human gait, but very heavily. The tracks were clearly visible, as the ground was wet. They were not at all like a bear, but more like human footprints, but longer, and the base was much wider. The heel was slightly wider than a human. "These creatures are called *tkis-katsi*, forest men. I have lived here for 20 years, and have heard that people used to meet them often before. Now, I do not hear about it any more."

District engineer B.F. Lobjanidze. In the summer of 1946 he was mowing grass with his brother and cousin, a boy, 3-4 km from the village. Suddenly the boy screamed and pointed to a person in a tree. The person lifted a hand to his eyes and looked at them, and then started to move from branch to branch, so quickly that the observers mistook him for a monkey. They came within 25-30 metres of the beast, and were very scared. "It was something terrible – not a monkey and not a human being. He looked more like a man of great height, but hairy on the head and body. Finally it hung from a branch, jumped down from a height of about 4 metres and disappeared into the bush."

Yaralova Robert, Armenian, 27, Lagodekhi district. About 6 in the evening, he was going with two others to the stream to fetch water. They saw a "wild man" completely overgrown with hair at a distance of about 20-30 metres. In the same year the boys went to the stream to bathe and saw another "wild man".

Several of the people from Lagodekhi talked about the possibility of taming the "wild man", especially the female, but that it is subject to certain rules of magic. Collective farmer Konstantin Begiashvili (Georgian, 65, Lagodekhi) said that during the lifetime of his father, a naked wild woman, "Ali" was caught, supposedly in a winter blizzard, where it was found sitting perched on the rump of his horse. According to the stories, "Ali" could not speak but would often laugh silently. She learned how to do simple errands for the family members. She could run very fast, and one day she simply left and went back to the forest.

An Susanova, 70, Tbilisi, told of an event he witnessed 60 years earlier. He was brought to the temple festival in Shavnabadsky monastery on the mountain 10 km from Tbilisi. After the service, a lot of pilgrims gathered in the meadow to eat. Suddenly someone cried "tkis-katsi"! Everyone turned and saw in the distance a humanoid creature, naked and overgrown with long hair. "He looked like a bear on its hind legs. He was heading in our direction with slow and uncertain steps. When he was still some distance away, he stopped and uttered inarticulate sounds. When one of the curious pilgrims started to walk towards him, he began to move quickly away. According to the stories of the pilgrims, the "forest man" was accustomed to get something from the table on such occasions. He was said to live in the dense forest, where he ate vegetation and roots, and sometime raw meat. Food was then laid out in a clearing some distance away, and finally the "man of the forest" went over there slowly, took the food and disappeared into the forest.

Dagestan
V.S. Karapetyan, who is a colonel in the medical service, has submitted an account to the Commission, which has already been described in the pages of the popular press, and has also attracted the attention of the foreign press. In October-December 1941 an infantry battalion, where Karapetyan was serving as medic, was transferred from Lankaran in Azerbaijan to the mountainous region of Dagestan. The event took place in the highlands near the city of Buynaksk in the cold winter season, near a village in the mountains. The local authorities asked the military doctor to examine a creature caught in the mountains a few days before, to establish whether it was a subversive dressed in a skin. This was of course in the first months of the Second World War. Karapetyan was shown into a cold barn, as according to the staff the creature could not be in a normal room because he sweated heavily. During its captivity the creature did not take any food or drink or say or ask for anything. The creature was a built like a human, a man, male, naked and barefoot. But on the chest, back and shoulders, the body was covered with dark brown hair. (It should be emphasised, that all the local people have strongly developed body hair.) This coat though was like a bear. The hairs were 2-3 cm in length. Below the chest the fur was thinner and more delicate. The hands were rough, with sparse hair on the back. The palms and soles of the feet were without hair. The hair on the head was very long, and reached the shoulders. The hair was very coarse and hard. The hair around the mouth was very fine. The man was standing perfectly straight. Height was above average, around 1.8 m. He was very large, broad and muscular. The fingers were very thick and strong and unusually large, the complexion was extremely dark, not human-like. The eyebrows were very thick. Under them the eyes were set very deeply in the head.

After this external description Karapetyan stressed that the creature, though so like a man in general build, was in fact an animal. He did not say anything, and had empty animal eyes. He blinked only rarely. All attempts to make him react to something, food, a hand, the sound of voices, gave no result.

Karapetyan only had a chance to observe the prisoner for a very short time, but did see something of great biological interest. His attention was drawn to the abundant lice in the chest, neck and facial region, especially in the eyebrows and around the mouth. These lice cannot be attributed to any known species of human louse. They are closer to a form of body lice, but of a larger size. It can thus be considered proven that Karapetyan did in fact see an animal, as animal do not carry parasites found on human skin.

In the summer of 1957, game warden V.K. Leontiev, who was surveying the territory of the Gutanskogo reserve, found tracks not belonging to either bear or any other known animal near a river. He heard loud cries at a distance of about 100-200 metres, and they were not from any wild animal or bird known to him. Finally late in the evening of 9 August, while sitting around the campfire, he saw a kaptar at a distance of no more than 50-60 metres. He looked at it for 5-7 minutes, until it crossed the snow, and then chased it for 8-10 minutes after a failed shot in the leg. This made the kaptar run up the hill with extreme speed. According to Leontiev's description, the kaptar was 2.2 – 2.3 m in height. The animal was very massive, with legs short and a little crooked, and long hands and fingers. All of the animal's body was covered with thick dark brown hair. The head was massive, and for a brief moment, when it turned and looked back, it was clear that it had a flat and human-like face. Summing up his experience Leontiev wrote: "If we compare the kaptar with a living creature, the closest match would be a very tall, massive and broad-shouldered man, who from head to toe was covered with long, thick hair… The kaptar is a peculiar creature, but it is unthinkable that he is just a beast, perhaps a special type of mountain monkey. The whole look of the kaptar indicates that this is a humanoid creature."

Leontiev did not have a camera, but he was able to sketch and measure the tracks. The length of the foot – 20 cm [*sic*], maximum width 15 cm. Four toes of equal length, about 5 cm, the first toe longer and wider, 9 x 3.5 cm. All the toes were clearly set apart. In general the trail of the kaptar is very different from bears, and also very different from humans. As for the cry of the animals, it was not comparable to anything the witness had ever heard. It was very loud, with a rhythmic repetition of low and high notes.

In July 1953, during a trip to Dagestan, K. Adibekyan and his driver E. Sahakyan were driving slowly on a deserted bumpy dirt road at dusk towards Derbent. Suddenly, a humanoid creature appeared in the middle of the road at a distance of 50-60 metres, and started jumping up and down and making strange sounds. "We got within 5-6 metres of the creature, and I was able to see it clearly. It was a man with long hair and no shoes, and he was naked. His chest was covered with thick grey hair. He was just over two metres, broad, with muscular arms." Despite the approaching car, the creature continued standing in the road, making strange leaps and shouts. With a sudden turn of the wheel, the driver went around the creature. When they looked back, the creature had not reacted, and was still jumping up and down. The next morning in Derbent Adibekyan told the local people about his strange meeting, and was surprised to be told that "such wild people sometimes appear in the Dagestan mountains. There is nothing to be afraid of."

Veterinarian Ramazan Omarov, Tlyaratinsky District. 20 August 1959. He was returning on a mountain path leading down from the upper village. It was about 8 p.m., and the visibility was excellent, about half a kilometre. "When I came to the great white stone (there is such a stone on the trail), I noticed an animal moving around at the bottom of the stone. I thought it was a bear and hid behind a bush. The animal, which appeared to be sitting, suddenly got up, and went on two legs towards me. It was like man and monkey at the same time. In my childhood I heard stories about the kaptar, but I had not believed them. But now I saw it. The fur was long and black like a goat. There was no neck, the head was sitting on the shoulders. The kaptar came quickly towards me, then just walked past. It was clearly a male. The head was long and pointed, conical or egg-shaped. The arms were long, and the hands almost reached the knees. Two hundred metres from me, the creature crossed the trail and sat down. After two or three minutes the kaptar got up and walked towards the hill. I was amazed to see him climbing up the hill very quickly with long strides. I thought of the chimpanzee in Tbilisi Zoo. The monkeys have shorter hair, rounder head, shorter arms and legs, and are smaller. The kaptar was at least 1.8 m, and looked more like a man than a monkey. It walked straight, and there was no tail. I could not see the ears, as they were covered by hair."

The views of local people about the "hairy wild man" can be divided in three groups. Some consider them evil spirits or a kind of ghost. Others believe that they are people running wild in the mountains, especially (according to one version) people who are infidels and do not follow the teachings of the Koran. Finally some see them as animals, animals that live on the edge of the eternal snow, among the rocks and in the forests. They are declining in number, and no longer approach the villages as they once did. They are generally described as sensitive, cunning and devious. They will sometimes attack people, or steal corn and potatoes in the fields. They walk on two legs, they are swift and nimble, and they are great rock-climbers. A few stories describe the differences between males and females, but we have very little information about young ones.

It can be very difficult to interview local people in Dagestan about these creatures. They generally avoid talking about the kaptar, and there are mystical and religious elements connected with the creature, obviously inspired by the Muslim mullahs. Young people can also be unwilling to talk about the kaptar for fear of being seen as uncultured and superstitious, so an interviewer must be very patient.

Chechen-Ingushetia and Kabardino-Balkaria

From Dagestan we move further north, to the territory of the Chechen-Ingush and Kabardino-Balkar Autonomous Republics.

In his work "The wild hairy forest man in the Main Caucasus Range", Professor A.A. Mashkovtsev writes: "In addition to the stories about wild hairy forest people of the North Caucasus we have found in older literature, we have also obtained about forty eyewitness reports from the Kabardino-Balkar Autonomous Soviet Socialist Republic and the Chechen-Ingush Autonomous Soviet Socialist Republic from people who claim that they have seen the wild hairy men with their own eyes." They are all from the years between 1850 and 1958. The accounts are very simple and prosaic, and there are no hints of embellishment and the like. The stories usually just describe a meeting with wild forest people, their behaviour and appearance. It is very interesting that the descriptions given by the Kabardians and Ingush are very similar to the descriptions of the "man of the forest" given by people from the Caspian and Black Sea coasts along the Caucasus. The wild hairy men are usually described as being very similar to humans, always walking on their hind legs, they are very tall, do not speak, their hands are very long, and the whole body is covered with hair. The wild forest people of the North Caucasus (*almas, almast, almasty*) live in remote and inaccessible places, but in the summer they can be found near villages in the melon fields and fruit gardens. Often you can meet wild people with young in abandoned sheds and houses.

Patlakh Suck, a resident of the village of Lezhi, claimed to have heard from a fellow countryman, an old man 140 years old *, that in the old days, when he was still young, sixteen hairy wild men came out of the reeds near the village of Ashluki Nazran. They would appear at night near a waterhole, close to the village. The local people tried to set a trap, and managed to capture one of the hairy people. He was shouting and roaring next morning. They all tried to explain to him that they meant no harm, but he did not understand human speech. When they unravelled him from the net he escaped and fled. After this incident all the wild men moved away from the village.

Kagermano Hassan told a story about his grandfather, who was hunting near the shore of the Assa River around 1850, where he killed a deer. Suddenly a "wild man – a*lmas*" appeared, apparently attracted by the dead deer, but he was eventually driven away towards the river.

Saadul Hasiev said that in the summer of 1912 or 1913 he was coming down from the mountain towards the Assinsky Gorge with his friends, when they stumbled upon a very decrepit humanoid creature lying face down on a rock. When they approached the creature, which turned out to be female, it just raised its head and upper body. The body was hairy, but there was no hair on the face. They could only see one or two large teeth in the open mouth. The men had a lot of trouble to control their horses, as they were very scared, so in the end they had to ride by.

In 1934 one of the residents of the hamlet of Bavli saw a woman overgrown with hair in the entrance to a cave in a remote area, and fired at her. He did not seem to hit her, and she disappeared into the cave.

Annecy Isaev said that in August 1940 he met a wild hairy almas in the road. It was screaming and howling and tried to attack him. When it attacked he fought it off with his knife, and it went away towards the Argun River. In the same area close to the Argun River another local resident met the same almas, and two rangers claimed that in 1939-1940 two civilians were attacked and beaten by almas in the same area.

An important piece of information is the fact that in this general area, where the front passed through twice during the Second World War, nothing has been heard of almas since. The ethnographer P.P. Bolychev has suggested that the thunder of guns and the passage of the armies forced the last of these creatures to move to some other area.

* *Translator's note: that's what the text says. Caucasians do sometimes claim to live to extraordinary ages.*

Based on the information collected by Bolychev, the almast are not human, but perhaps apes. They used to live in the wooded mountainous parts of Kabardia and Balkaria. They had special dens, but would sometimes sleep in abandoned huts on alpine pastures. Some people had seen them eating corn, onions and carrots, as well as the many edible plants found in the forests of Kabardia from spring to autumn, wild fruits, nuts and berries . One old man claimed that they would also hunt and eat small birds. The almast used no clothes. Their body is covered with thick black or brown hair, except the breasts of females. They are black, and hang down, when the animals are running. The hair on the head is long and wavy, and some people claim they have red eyes. The almast move very quickly. Not even horses can keep up with them. Horses are afraid of them, and they hate dogs. There are lots of stories about almast being chased by dogs. They are nocturnal, and only appear during daytime if they are scared or chased out of their hiding place.

Bolychev has suggested a nationwide collection of specific data on the almast, and said that even if only a few specimens are found alive, a park should be established to save them from extinction. He emphasises that the collection of data in the Kabardino -Balkaria area is difficult, as some old people are not willing to talk to outsiders.

Here are some survey data reported by R.D. Varkasovym. In 1956, near the caves in the area called Kardahurt, a group of men were terrified by a creature that some called *gubganana* ("old field"), others almast, but their descriptions were the same: build as a man of medium height, a hairless human face, with long hair down to the shoulders, and a body covered with a red-brownish hair. In the summer of 1957, about two kilometres above the same place, the farmer Afaunov Arba said he saw an almast in the valley of a nameless river, about thirty feet away. According to Galim Evgazhukova from the village of Kizburun, from 1936-1938 he saw many almast, both male and female, in the western part of Kabardia. They live in numerous caves, but they leave when the collective graze their herd in the area. The best time to look for them is from May to mid-June. He had seen a female even when he was a young boy. They are not afraid of single people, but they avoid groups of people, and they are very much afraid of dogs. They cannot speak, but they make a sound like a human shouting. They are strong runners.

Yusup Arutyunov, 35 years old, collected many stories about almast living in the nearby wilderness in his native village of Zhemtala. He said that before the war they were not rare, but after the war only very few people have seen them. In 1954, Ishbuhov Rashid, 22, was stealing apples in another garden, when he came face to face with an almast. It was said to be wholly tame and fed by the owner of the garden. Talib Balkaria, 40, the farm shepherd, is said to have seen an almast with a baby in its hands in the mountains near the village of Lesquin, not long ago. Several other people in the area also claim to have had brief encounters with almast.

R.D. Varkvasov has said that the local people absolutely believe in the authenticity of this almast, and points out that, before the war, the humanoid creatures were not uncommon. In the winter they probably moved to warmer places, as all the meetings took place in the spring or summer. According to local folk tales the almast feed on all kinds of small animals, wild apples, pears, nuts, etc., and when they live around settlements also eat fruit, corn, bread, cheese, raw and cooked meat, if they can steal it from the local people. Many of the stories emphasise that they stink quite badly. The smell is very unpleasant for humans. The report also contains information on individual families who have kept tame almast or taught almast to come for food at night.

In 1918 a farmer named H.N. Akhmetov, was spending the night with a friend in an old shed near a pasture in the Psyarisha area. They had left a smouldering fire in one corner of the shed. During the night they woke up and saw a stranger by the fire. It was taller than a human, was hairy all over, and "his hands were hanging down to his knees". The stranger mumbled something and then left the hut, and when the dog chased him he ran away with a terrible cry.

H.G. Thagapsoeva has summarised data collected on the appearance of the almast. The body resembles a human. The neck is thick and short, and the entire body is covered with hair. The jaws are powerful, and the teeth are large. Many have noted that the animals mostly sit on their heels. There is a clear difference in appearance between the sexes. The females are shorter, have very long hair on the head and a more female figure with well-developed breasts. Several Kabardian people have reported seeing young almast, but no one has ever reported seeing caves or huts belonging to the *almasty*. They have been seen in abandoned farm buildings and remote huts and yurts. They are clearly not able to build their own home. The almasty eat plant food, and they have attacked plantations. They are not predatory, i.e. they do not attack man or his animals. They are basically timid and defensive, and they are able to move very quickly.

Finally Thagapsoeva has demonstrated the existence of a peculiar Muslim superstition regarding the almas. According to some, people who kill or offend an almas cannot escape God's punishment. Thus many elderly people revere almas, and are ready to provide them with any services hoping that God will reward them in the afterlife.

According to H.C. Getezheva, in 1919, he witnessed how people armed with anything they could lay their hands on ran to the forest and started beating a very scary-looking woman covered in black hair, and lying on the ground. She looked to be 45-50 years old. He was told she was taken away into one of the houses. In 1926, the same informant saw a very scary hairy woman following his brother around. She had reddish eyes, was rather short, stocky and had short legs. The head was covered with thick tangled black hair reaching the shoulders. She had strongly developed hanging breasts. She was holding her child, about one or two years old, in one hand, and the informant's brother in the other hand.

In July 1951 G.E. Tokhov was in a deserted valley at the confluence of the Greater and Lesser Nalchik Rivers. He went into an abandoned house, and saw a dark creature in a doorway. "I knew from stories the old people had told that this was an almasty". First he wanted to get closer, but changed his mind, went round the building and found a hole to look through. "Right in front of me in the corner sat a creature resembling a human, hugging its legs and with the head resting on the knees. When it heard the sound of my footsteps, it lifted its head and began to listen. Then it leaned forwards on its hand, and rose. Now I could see it clearly. The body was covered with fine hair, and there was long, thick black hair on the head, that reached down to the waist. The eyes were big and red. It was about 1.8 m, with long arms and hands. The legs were short. It was a female with breasts hanging down almost to its waist." Suddenly the creature heard the sound of the horse, and became more alert. This frightened the observer, who jumped on his horse and rode away.

[Translator's note: the next paragraph is a jumble of short notes about a man apparently getting into a fight with a hairy creature, but due to lack of coherence and specific data, I have omitted it.]

About 30 years ago in the village of Argudan (Leskensky District), a woman came out on a cold winter morning and went to a nearby hut with a deep pit for garbage. "Looking into the pit, I screamed with fright. In the pit was a naked hairy woman with a little hairy baby, that she tried to wrap in a piece of dirty cloth." The woman threw down a hairy old blanket, hoping to ward off evil, but the hairy woman ran off clutching her child.

An employee at the farm in Urkutin Baksan has claimed in a letter that all people, young and old, in Kabardia are aware of the existence of the almast. Then he gives a list of villagers who have personally observed the creature: twelve men in total. Hamid Sintashev, 50, 20-25 years ago saw a "girl" almast caught in a garden where she came for the apples. She was held overnight in an empty room and released in the morning. "She could not speak, but was crying and whining all night." Hamid Pashtov, 50, saw an almast in a haystack in 1953. In 1952, Mohamed Kardan, 25, met an almast in a cornfield at a distance of 2 - 3 metres. In December 1953 Tengiz Aves was a watchman at the mill in the Cuba area. "I heard something on the second floor of the mill, and when I went up there, I saw an almast eating handfuls of corn. It promptly ran away screaming." In the summer of 1954, two almast were found in a field in the same area. One was sleeping, and the other one was eating something. According to a correspondent called Talunew, it has now become more difficult to see almastys. It is easier in the summer, when they come down into the valley. They eat mainly summer corn and fruits.

A very interesting story was handed to Auldin Elmesov, who is a researcher at the

Mountain Geophysical Institute of the USSR on Mount Elbrus. It was told by his uncle Magil Elmesov. In 1938 or 1939 Magil Elmesov worked as a collective farm herdsman. Every summer in June and July he climbed the Baksan Valley with the collective farm animals, usually horses, and put them on pasture. Every summer, Magil also brought out his apiary, and one belonging to a Russian bee-keeper from Novo-Ivanovka near Nalchik. When Magil Elmesov went to visit him the bee-keeper said that, just seven days earlier, his younger brother (or possibly son) had shot a creature very similar to a man because it was robbing him of honey. The bee-keeper led Magil Elmesov into the bushes about 150 metres from his tent, and showed him the dead creature or “Satan”. Magil never forgot what he saw. The already decomposing corpse was very similar to a human. It was hairy, and the face was elongated like a dog's. The limbs were very long, and so were the fingers, but the hands had no hair. He immediately realised that this was an almast like the ones he had heard of.

In the summer of 1960, Professor A.A. Mashkovtsev made a reconnaissance trip to the Kabardino-Balkaria area, where he conducted extensive work in a number of villages, and also studied zoology and topography.

First of all, Mashkovtsev tried to ascertain the attitude of the local population towards the almast. To his surprise, the children and women simply pointed to a bushy area near the village, and said he could find almas in there. A group of women even asked for his help in scaring the almast away from their land, because they were eating the ripe fruits and corn. When he was speaking to middle-aged and elderly Kabardian men, Mashkovtsev would ask a cautious question: whether all the stories about the almast were pure superstition, a kind of belief in the devil. The answers were always the same. The devil was a disembodied spirit, but the almasty were material beings. There was an almasty female and a male, they would have young, gradually grow old and grey, and die of old age and illness. There are many cases of sheepdogs that have chased and bitten almasty to death. In the old days, there were several cases of almastys being domesticated and living for several years in Kabardian families performing very simple tasks. There were also supposedly several cases of sexual relations between lonely and solitary male Kabardians and almasty females. Since the establishment of Soviet power, residents say that the number of almasty has been sharply reduced. However, they still meet almasty every year in the surrounding areas. There are more of them to the east, which has more lowland forests than in the western part of Kabardia. In general all of the rural population in the surveyed villages in Kabardia (Baksanenok, Psinshoka, Kurkuzhin and Nartan) believe in the existence of humanoids completely covered in dark hair. They do not speak or use tools, but they know how to throw a stick, clods of earth or stones. They only very rarely attack humans. They are peaceful, and will try to communicate, especially with women. According to the Kabardians, they often meet almasty in farms or in gardens. In the fields in the summer, the almasty will appear around field camps where food is prepared, and eat the leftovers.

H.K. Getezhev, a Kabardian, told a story about how his brother had brought a hairy woman with a child he had found in the sheepfold. The same wild woman came to his hut in a cornfield several times during the summer to look for food. "It is not very afraid of people, but it is cautious, so it usually appears when I am alone."

H.G. Gutova pointed out a vacant lot between the estates, overgrown with weeds, where she saw a black hairy humanoid in 1957. The neighbours added that in the same weeds they saw a female almasty with a child in the summer of 1960.

An old man named Khizir Gubashev, 80, said that he herded cattle on the mountain pastures for 20 years and only once in that time met an almasty, but down on the plains he met an almasty twice - in 1936 and 1958. "One morning I went to the river, and suddenly saw an almasty: it was a tall wild man whose body was covered with black hair. I saw him very close, perhaps 15 – 20 steps away, and noticed that his eyes were elongated and placed obliquely like the Chinese. The fingers were thick." But Gubashev also added that he looked as little as possible at the almasty, as people get sick from it, and he was still ill 3-4 months after the meeting. Gubashev also said that a few years ago the shepherd's dogs caught and strangled an almasty in the elm forest. The body was left in the forest for a long time until it decomposed. Although his stories are very realistic, Gubashev considers the almasty as being demons punished by Allah for serious infractions, thrown to Earth, turned mortal and with a disgusting look, resembling both an animal and a person.

Elena G. Aleksanova, 63, said that in June 1959, about 11 a.m., she heard meowing sounds and found a child about 6-7 years old sitting among the potatoes. His whole body was covered with black hair, his hair was dishevelled, and the skin of his face was completely black. She also added, that each year she hears stories from others about almasty in the woods, in the bushes and in the cornfields. Mashkovtsev also added that he heard a story from the area, that a man would feed an almasty that came to the estate, but that he also tried to hide it from his neighbours.

The overseer at the Gedukinskogo Reserve, Salim Zagashtukov, said that in July 1946 he left their hominy in a small abandoned house on a meadow, and went to mow the grass. When he came back, he surprised an almasty eating the hominy. It jumped up at once and ran into the woods. He saw it again in the forest later that summer.

In 1959, in the mountains near the Chegem forest and the Inal Dag mountains, there was a great tragedy. Two shepherds saw an animal down in a wooded gorge. One of the shepherds ran with a gun in that direction. Suddenly they heard a terrible scream, and the other shepherd ran to help with his gun. Neither of them returned. They were found two days later, terribly mutilated with severed hands and heads. The guns were found with their barrels bent. The head of the local pig farm, Nashiri Panzhokov, said that "I think the shepherds were killed by a huge wild mountain man, something like a gorilla. We know that, apart from the harmless almasty, there are also scary and very fierce wild mountain men that look like gorillas."

R.D. Varkasovym recorded an observation by Khamila Tlenkopacheva. In 1937, in the camp, where he was secretary of the Komsomol organization, someone stole the meat at night. The following night at midnight a watchman saw a humanoid creature that had very long arms, narrow shoulders and elongated head. The guard raised the alarm, and all ran for almasty, but failed to catch him.

Agronomy High School Teacher G. Tettsoev, from the village of Baksanenok Upon, told Mashkovtsev on 31 July 1960 about the events that took place two days before, the night of 29 July, at sowing mill №2, where he spent two days with a team of high school students. When they went to sleep, the dogs were barking at something in the cornfield, and were excited for a long time. At about 2 in the morning a frightened horse galloped through the field. Then came cries of alarm from the house where the girls slept. One of them, L. Zarakusheva, said she slept on the doorstep, and was awakened by the rustling. She then woke her friend, and both looked in horror from under the blanket at a humanoid creature who came to the door, squatted down, and looked at the sleeping students. It was a huge and very broad almasty. Hearing the voice of somebody shouting at the frightened horse, the almasty slowly got to his feet and walked out of the house. In the morning Tettsoev questioned the local guard, who replied that every summer they would have almasty living in the fields near №2 field camp. And that this was the third time an almasty had come to the house that summer. Mashkovtsev visited the field camp on 4 August 1960, and talked to the watchman and two cooks who cook food for the farmers. The watchman said that visitors would see an almasty if they spent just a few nights at the field camp. But the cook was angry with anyone coming to catch the almasty. He said it would mean disaster if it was caught. Mashkovtsev left with the impression that sometimes whole families of almasty can turn up at the camp.

In the summer of 1959 a young Kabardian caught a small female almasty aged 7-8 years. All the residents ran to the outskirts of the village, where he was holding the hand of a wild forest girl, who was covering her face with the other hand. The other villagers immediately ordered the boy to take her back to the forest where he caught her, and release her into the wild to avoid the accidents that would otherwise befall the village. Many people claim that they have met with misfortune if they have hurt or harmed almasty in any way. This makes it difficult to study the subject, as many people will not talk about it.

Mashkovtsev provides convincing evidence for the fact that the almasty in some way is under the protection and patronage of the Koran. Caucasian Muslims, he says, will obstruct our search for these creatures.

By analysing all the collected records, Mashkovtsev comes to the conclusion that they are simple, sincere and realistic: “I have to admit that it is possible that some kind of humanoid creatures, maybe Pithecanthropus or a Neanderthaloid type, still inhabit the territory of Kabardino-Balkaria and Chechnya-Ingushetia in very small numbers. In the summer, they are found on the flat steppe territory of Kabardino-Balkaria. Some of them probably winter in the forest-steppe around Kabardia, sheltering from the weather in empty human buildings. However, not all almasty in Kabardino-Balkaria are 'freeloaders' and live near human settlements on the plain. Part of the 'wild forest hairy people' are living in caves in the mountains near Elbrus, Mount Inal Dag."

Are the mountain wild men the same as the ones living on the Kabardian plain? Mashkovtsev puts forward two possible solutions: 1) they are two different types, two special races, 2) the wild "giants" of the Central Caucasus, that reach enormous size and great strength, are simply very large male specimens of the same wild mountain people.

With regards to the almasty living on the plains of Kabardia, Mashkovtsev makes the assumption that they are mostly female individuals, especially when they are with young or pregnant. They are forced to come in close contact with people to get food, so they steal vegetables and fruits from the gardens, eating leftovers, corn, melons and berries in the field.

Finally, we conclude our review of data with excerpts from the records of S.G. Muguet. Sawyer Hoza Zhankiraev of Baksan met a hairy man in the woods in the area around Elbrus. The hairy man grabbed him by the arm, but Hoza managed to free himself, and the wild man did not pursue him. Engineer I.I. Simonov, who in 1956 worked near the village of Zhemgala, said that the residents of this village once came to work one morning and caught a wild man near the extinguished campfire. According to A.K. Tembotova, partisans hiding in the ravine area near Keech-Malka during the last war for several days fed two almasty who lived in a nearby cave. In 1935 (or 1937) in the village of Barazbi, Nogmov Kamenomost who guarded a warehouse was approached by almasty, but he was so frightened he could not shoot even if he wanted to. In 1939, a tractor driver named Tokhov was returning home through a sunflower field in the evening, when he met an almasty only three or four steps away. Tokhov lit a kerosene soaked piece of hemp which he carried with him at night to guard against wolves. In the light he could see the almasty was the size of a normal man, but covered with hair. In 1938 or 1940, a watchman called Mato Zherikov who was chopping firewood saw an almasty near a field camp. Later he returned with his nephew, and they found the almasty in the same place. When it saw them, it got up and left. Another night in 1938 or 1940, Zherikov passed a toilet which was surrounded by barking dogs. Zherikov threw a large stone down into the toilet (these are often without roofs), and in the next moment an almasty jumped out, ran away and hid in the weeds near the nearby cemetery. In 1936 or 1937, a shepherd near Aursend heard a strange cry, and eventually saw three figures in a cave. He ran back to the village of Kamenomost, and brought about 30 people back with him. In the cave there were three almasty with their backs to the people. They were making muttering sounds. Since the villagers were standing in the opening of the cave, the almasty did not try to escape. The villagers could study them from a distance of only 3-5 metres. They were a little shorter than humans, stocky and covered with black or brown fur. They gave off a sharp and unpleasant smell. They had no tail, slanting reddish eyes, and a head slightly longer than humans. Their heads were covered with very long, dirty and matted hair. The hair on the body was not long, but dense. The chest hairs were longer than on the sides and legs. Local people are certain these beasts can still be seen, although they occur much less frequently than before.

Shahen Mohammed from the village of Kamenomost claims that on a spring morning in 1933 he was attacked by an almasty. There was a struggle, but the noise alerted his dogs who chased the almasty away. It ran off and hid in a cave. In the same cave in February 1960 a hunter named M.M. Balaga, who was chasing a wounded fox, ran into an almasty - a naked man covered with brown thick hair. The hunter left the fox and ran home. One night in the summer of 1945 or 1946, Murad Hofiyugov, who was riding on oxen in the Psybundzh area, met two almasty on the road. Teak Shabotekov, an old man over 80, said that until the last war he had seen almasty many times near Kamenomost at night. But after the war the almasty have disappeared from these areas. 105 year old Bestaev Shanahov from the same area claimed that when he was about 50 years old he had seen an almasty in a cave near the Malka River district. It was the height of a small man, with red eyes and covered with black shaggy hair. It was very long on the head. It had a very unpleasant smell. It walked almost like a human. Shanahov and the almasty faced each other for a few minutes, and then both retreated. Shanahov claims that up to this personal meeting he did not believe the stories about the existence of the almasty.

In some cases eyewitnesses claim to have seen almasty wearing what looked like loin cloths or some form of cloth tied around their waist. According to villagers they would not only steal food but also clothes.

According to the records of S.G. Muguet, some Kabardians distinguish between two varieties of wild men:

1) *almasty* - long hair, eyes with vertical pupils, they live close to the villages, are scavengers, stealing corn and other agricultural crops in the fields;
2) *mozyl* – this forest man is found away from the villages, short hair, chest keeled, eyes with a horizontal pupil.

The following was told by L.S. Ozrokovoy who is a graduate from the Leningrad Conservatory. “In the past before the war, our country had a lot of almasty. Many people have seen them. When my mother was young, she saw them too. She had been working in a field with other farmers, when they suddenly were approached by a group of almasty that started copying their movements. The farmers were frightened, but an old man named Ali, who was sharpening the hoes because he could not work in the field said 'Do not be afraid, they will not touch you.' When we were little, and went into the woods, my mother always warned: 'If you see an almasty, do not tease it or throw stones and sticks.' My father was a great hunter, and hunted a lot in the mountains and in the forest. He met the almasty many times. The male is called *mozyl* and the female *gubganana*. After the war the almasty has been seen much less. "

Shamil Ozrokovoy, 70, told how 40 years ago he was awakened by barking dogs at night. When he looked out, he saw something in the corner of the garden. "I saw that it was a *gubganana*. It was just like normal people, only very tall, and covered with hair. It was very long on the head. The breasts are very long. The face was terrible. It looked like a human, but the lips were pushed forward like a monkey." The dogs were afraid to get close, but Shamil picked up a big stick, and went over to the gubganana and hit her. She screamed very loudly and jumped over the fence. It was almost two metres in height. "Before the war, there were a lot of almasty. Now very few are heard of. They eat everything that grows in the gardens: cucumbers, tomatoes, corn on the cob. In winter they live in the forest. "

Zhandarbi Dzahmisheva, 59 years: "I saw an almasty in July 1937 on the southern slopes of the Western Dag, more precisely, on the eastern shore of Tyzyl. I was passing a flock of sheep, which walked along the shore, and I was above them on the hillside. It was 10 - 11 o'clock in the morning, and the weather was good. When I came around a corner, I suddenly saw an almasty 15 metres away, if not closer. He was sitting in the sun. I did not have a gun, but I stopped and whistled. When he heard the whistle, he turned to face me. He had long hair on the head, and seemed almost human. When he got to his feet, I could see he was hairy like a buffalo. The he walked on two feet into the cave he was sitting in front of. He smelled very bad. The tracks looked like human feet, but were very flat. The cave is very deep. In the 10 years I worked there, I only saw the almasty once. But every year I heard about three or four people having seen it. "

Erzhib Koshokoev,70 years, the village of Old Cherek, said that before the war there were a lot of almasty in their territories, but now they are very rare. Personally, he saw almasty twice in his life. The first time was in September 1944 when he was part of a detachment of soldiers policing an area near the Black River. The man in front of him suddenly stopped his horse and said: "Look, almasty!". A few metres in front of them, there was a gubganana eating hemp seed heads. When she saw the soldiers she ran quickly on two legs and hid in a nearby hut. The troop dismounted and surrounded the hut. The gubganana was jumping back and forth and muttering and making noises. It was very agitated. "When we got very close, she suddenly gave a terrible cry and rushed straight at the people. She ran as fast as a horse. Everybody was very confused, so she broke through our chain. She then jumped into a ravine and disappeared into the bush surrounding the river. She was about 1.8 m in height and the whole body was covered with red hair, almost like buffalo hair. I also went to look at the tracks in the ravine. They were very small, and I was very surprised."

The second time Koshokoev saw an almasty was in the summer of 1946, early in the morning, when he was riding along the edge of a cornfield. The almasty was sitting beside the road, but when the rider approached, it got up and disappeared into the corn. According to Koshokoev hemp is the favourite food of the almasty, so one should track it at night near a hemp field. It makes a lot of noise when it is eating at night. Almasty also like watermelon. They can do a lot of damage on a plantation.

Ibrahim Guliyev, Balkar, 31, recalls that when he was 7 - 8 years old, he was searching for a lost goat, when he came across an almast lying curled up in the grass. It was covered with dark reddish hair, and it was sleeping. It was the size of a girl about 13 years of age. Some of the reports concerning such random encounters are very recent.

Balkar Salim, 39 years, from the same area, reported that in 1959 something scared his horse, and he jumped outside with a gun. "I saw a man approaching me. I called out to him, but he did not answer. He was grinning and laughing like a hysterical woman. I pointed the gun at him and shouted: 'I'll shoot!', but he just laughed even louder. I lit a fire and realised it was an almasty. When he saw the fire, he began to back away and left."

Nazir Guliyev informed us that in October 1961 a hunter from Tyrnov Auza who went down to the river returned very quickly. He was very excited. He had seen a humanoid creature in the gorge near the Devil's Pit. The creature was sitting on his haunches. His arms were so long that the forearms were lying on the ground. It was a male, with no clothes, and covered in grey hair. He was very thin. He was either sick or very old.

In the autumn of 1961 Khazizov Bechukuevym, 19, had the following experience: "I tend the horses near the central manor farm in Elbrus. On a cold foggy night we were sitting around the campfire, when a human figure appeared out of the mist. When it got closer, we saw that it was an almasty. It sat down by the fire in front of us, and kept looking at us. It was a male. The hair on the head was very long, and the body was covered with black fur. It kept coming back for three nights. On the second night, we tied the riding horses near the fire. The almasty came back to our fire and sat down. The same thing happened on the third night, and we heard several stories of how the almasty would come to the fire of the shepherds when the weather was bad."
On 21 July 1962, Nina Bodaeva, 19, said that two weeks earlier, when she was working in a maize field near Psyho, she went to get water in the river around noon. When she got closer she saw a man sitting under the fence near the watermelons on his heels. He was covered all over in red hair. The girl was very frightened, threw her bucket at the creature and ran away.

There are also a few cases of almasty being captured or domesticated. Abuzer Frill, 28, has told that in 1939 his father was putting a horse in a stable, and saw a man hiding in a corner. He called out to him, but the man rushed to the door and jumped out of the stable. His father chased him on horseback, and saw that it was an almasty covered in hair. The hair on its head was long enough to reach the ground. The almasty then ran into a house, and his father ran after it and grabbed its hair. Several people came to his help, and helped Frill's father to hold him. The almasty was locked in a barn and kept for 2-3 hours. After that, the people decided to send him to Nalchik. The father received an award of 500 roubles for the capture.

In the summer of 1962 Chariton Bekulov was able to capture a *mozyl* – a humanoid creature covered in thick black or brown hair. His father, Zachi Bekulov confirmed the story, and said his son received 500 roubles.

S.T. Shtymov, 45 years, Ph.D, director of a high school, said that his brother had almost captured an almasty. "It is true that there are almasty living in our country. I saw one as a child. It was the summer of 1924 in the second half of August. We were driving towards the city of Cool. We stopped, and I was left near the carts with the horses. I started to wander around, and went into an old abandoned hut. It was early afternoon, and after having been out in the sunlight, I did not notice anything. Then, as my eyes adjusted to the darkness, I saw a hairy creature sitting in a corner. I was standing in the doorway, blocking the exit. I screamed in fright. The cry brought my older brother Alexander, who was then 40 years old. He saw that it was an almasty, as he had often met them. He told me: 'Do not be afraid, do not pay attention. It is a harmless creature, you see – it is almost the same as we are'. When my brother came closer, the almasty made a loud squeak like a monkey. My brother tried to grab him, but the almasty resisted. Finally the almasty broke free and ran out of the barn, where he jumped into the thicket and disappeared. The hair on the almasty was thick and black. The hair on his head was very long. I did not see his eyes because of the hair. He was rather small in height, maybe 1.30 – 1.35 m. In 1925, when my brother was the chairman of the village council, a doctor from Baku and Yerevan came to see him. The doctor was interested in the almasty. He made a special trip to see my brother, because his own brother, who had been a guerrilla fighter in the civil war, had seen these creatures several times. After that they went out to search the area where we had seen the almasty. They went around the top of the entire area, and into the abandoned sheds. Finally, one day, in the same area where I saw the almasty, my brother came upon an almasty in a shed. He rushed at him, shouting that he needed help, but before the doctor had come down from his horse, the almasty escaped and fled."

One of the revered elders among the Balkars, Thapa Hazievich Shavaev, about 100 years old, told a very curious story about the almasty. "In the old days there were a lot of them, and they often lived with humans. In our village there was the Ahmatova family. They always had almasty living in their house. I often slept in the house. Every time the family sat down for lunch or dinner, the almasty had their rations in the courtyard. One time they forgot to feed the almasty, and when we returned, he had searched the whole house for his rations, all the dishes, all the pots, spoons, bowls etc. In the past you could always find an almasty. Now it is very difficult."

All of these sightings are from Kabardino-Balkaria. We can see two important types of information in them. Firstly there is a lot about young creatures, which would suggest that the Greater Caucasus is an important breeding area. Secondly, there are a lot of cases of domestication or semi-domestication of these creatures.

Svaneti, Abkhazia

G.T. Gvalia has summarised stories from old people about the "wild man" (men, women, adolescents) in the forests of the river district of Tskhenistsqali (the Svaneti Range). "It was said that the wild men are very similar to humans in height or slightly taller; they are naked, but covered with hair; able to run fast, and swim. They could not speak, but could cry very loudly. There were also very small "wild men", about the size of a 10-year old boy. It was impossible to catch them, because they were very clever and could run very fast."

In December 1961 S.V. Muguet and K.M. Velikanova made a small study of this area. Here are some of their records. Four separate people said that Anton Naveriani, who lived in Gvebrya, and died in 1960, was praying in a monastery, when he met with a naked woman with long hair and large teeth. He pulled out his dagger to defend himself, but the woman screamed and hid in the bushes. A patient of doctor Nijeradze named Dovletkhan and his daughter said that 30 years ago they had seen a hairy girl swinging in the branches of a tree.

20-30 years ago, in Bechu, a 5 year old girl called Katya Hurgani went missing during harvest. After a day's searching she was found high in the mountains in a remote canyon. The girl said she was kidnapped by a hairy woman, who dragged her up in the mountains, but that the woman ran away when she heard the voices of the people looking for the girl.

According to the Abkhazian stories, the *ocho kochi* are slightly taller than humans. They are covered with hair, have large teeth that look like human teeth, but more spread out fingers, especially the thumb. They look a bit like bears, but hunters know them well, and know the difference between them. Some stories suggest some form of interbreeding with humans. When a child is very ugly at birth, the people decide the father is an ocho kochi.

In Svaneti, in the village of Ptischi, about ten years ago, an ocho kochi got in the habit of stealing honey. The residents of the village took honey and mixed it with vodka and used it to drug the ocho kochi. After that they tied it up, and sent it to Sukhumi. What happened to it then is unknown. Staff members of the Sukhumi monkey sanctuary had heard that a wild woman had been caught in the mountains ten years ago, but had heard it was sent to Moscow, not to Sukhumi.

In the Karachay region people talk about the *arach-kschi*. It looks like a naked man wearing a coat, but it is taller than a man. Sometimes it attacks people, or at least the local people were afraid of it, and would only walk into the mountains with a gun. Some people think that along with the "real" wild people are Karachay people living wild. In the summer of 1962 I gathered some additional information about the Svaneti area. The old people told me that about 20 km away it was possible to find *devas* – hairy wild men. The villagers in this area are also afraid of walking on the roads at night, out of fear of these creatures.

In Abkhazia there are a lot of folk tales about the *abnauayu* ("forest people") that used to live there. They are strong, ugly and hairy. Anthropologist G.F. Chursin has written that: "I have collected lots of detailed information about the abnauayu and its way of life. They are tall and completely covered with hair. The abnauayu hunt at night. In the daytime they hide in dense forests. In the old days, when there were more forests, the abnauayu were common, but when the forests are cut down, the abnauayu will be killed by the hunters and shepherds in the area."

Dzhalmat Dergolia, 88 years, told a story he had heard from his grandfather, which means the events must have taken place no more than 200 years ago."In those days nobody went upstream on the Dzhuya river, because this was where the abnauayu (the forest people) or the *adou* (wild men) lived. It was impossible to catch the abnauayu. But a well-known hunter Mardipo Mardas managed to do it by pouring vodka in the water they used to drink. He caught three. All three were tied up and taken to Prince Shervashidze (ruler of Abkhazia and Samegrelo)." All three were male. They looked like a man without clothes, but they were completely covered with black hair ("like a lamb"). Some say they were up to three metres tall. When they were living with the prince, they were fed meat. They lived there for about a year, but the fate of them is unknown.

[Translator's note: This chapter ends with one page of discussion of Neanderthal fossils found in Asia. Today our knowledge of Neanderthal fossils is far greater, and considerably more fossils have been found, so I have omitted this page, and would guide interested readers towards more modern discussions of Asian Neanderthals.]

Baksan Valley: the Baksan Valley in Kabardino-Balkaria. Chapter 9.

Some members of the Commission for the study of the Snowman (from left to right), Prof. B.F Porshnev, Prof. A.A. Mashkovtsev Prof. P.P. Smolin, Dmitri Bayanov and Doctor Marie-Jeanne Kofman. Chapter 1.

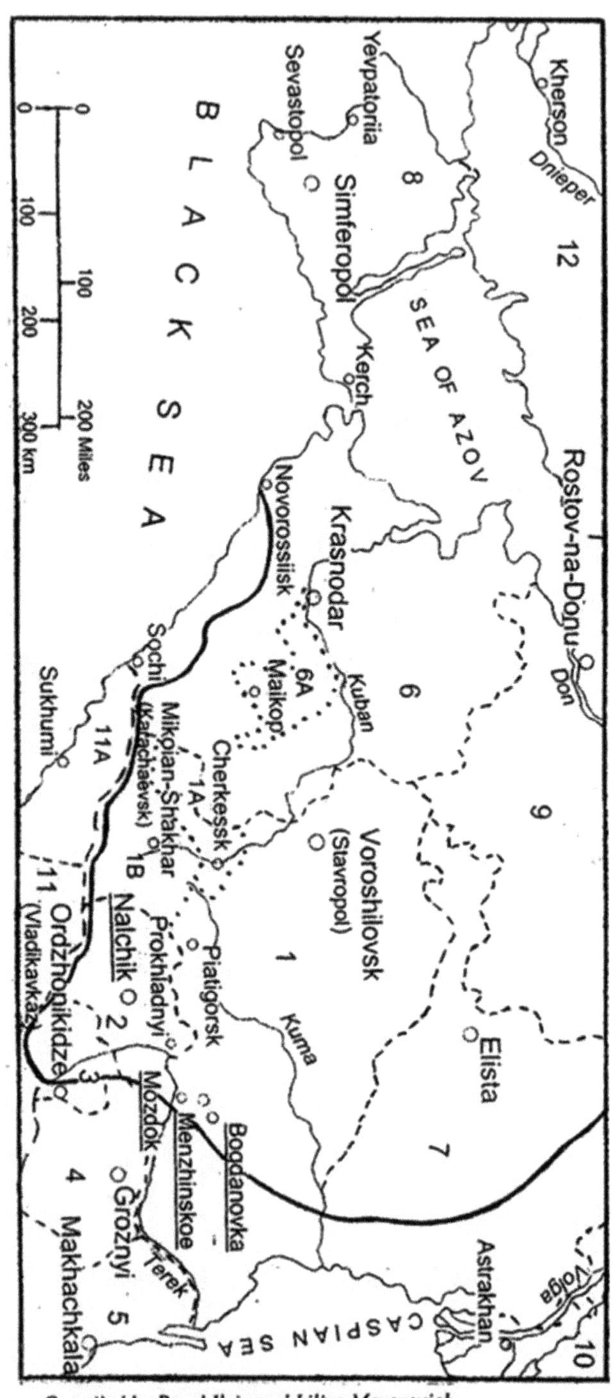

Compiled by Pavel Ilyin and Liliya Meyerovich

German advance to Caucasus: The black line shows the limit of the German advance into the Caucasus in 1942-43. Region 2 is Kabardino-Balkaria, where any almas would presumably have been driven out. Chapter 9

A 'yeti' scalp worn by lamas at the Khumjung Monastery in Nepal. This was the one borrowed by the Hillary expedition.
Chapter 4.

The ksy-gyik's sleeping position according to Khakhlov.
Chapter 3.

The Gobi desert in Mongolia.
Chapter 6

Mount Everest.
Chapter 7.

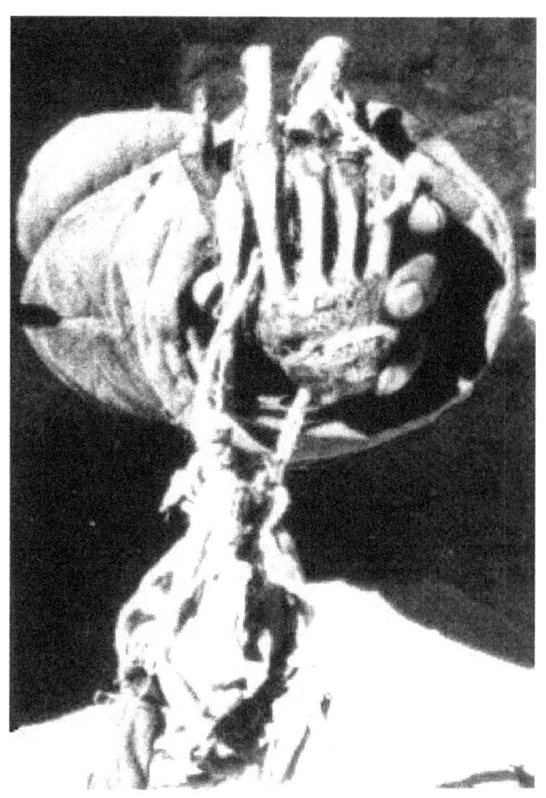

The hand at Pangboche monastery
photographed by Peter Byrne in 1958.
Chapter 11

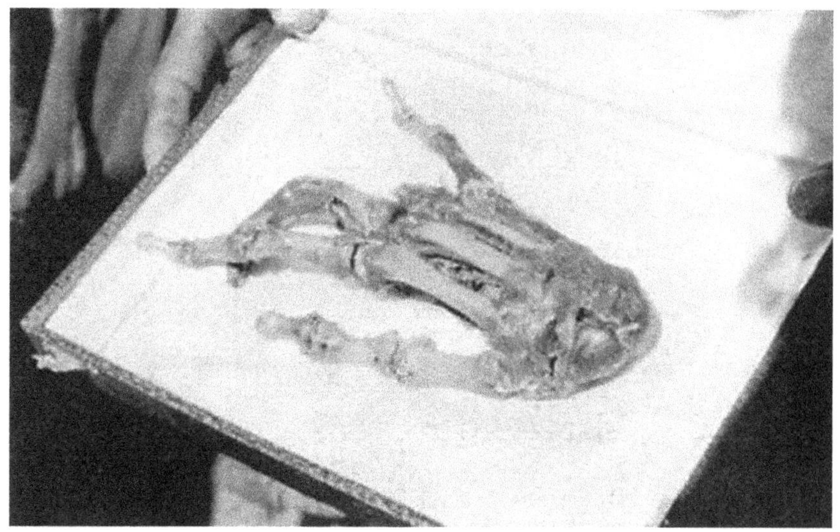

The hand at Pangboche monastery photographed by Professor Teizo Ogawa in 1960, after the partial substitution by Peter Byrne.
Chapter 11,

Distant view of Mount Elbrus in Kabardino-Balkaria.
Chapter 9.

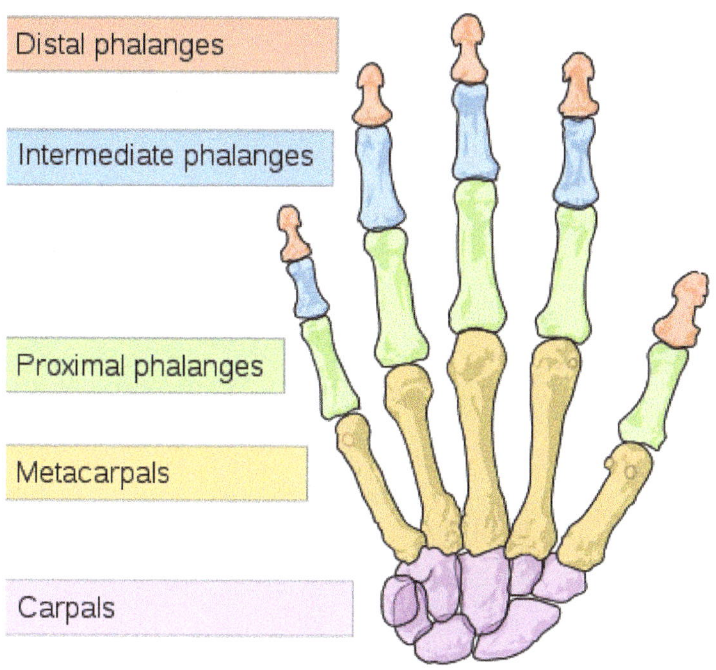

A diagram of the human hand with the bones identified, to illustrate the discussion of the Pangboche hand.
Chapter 11

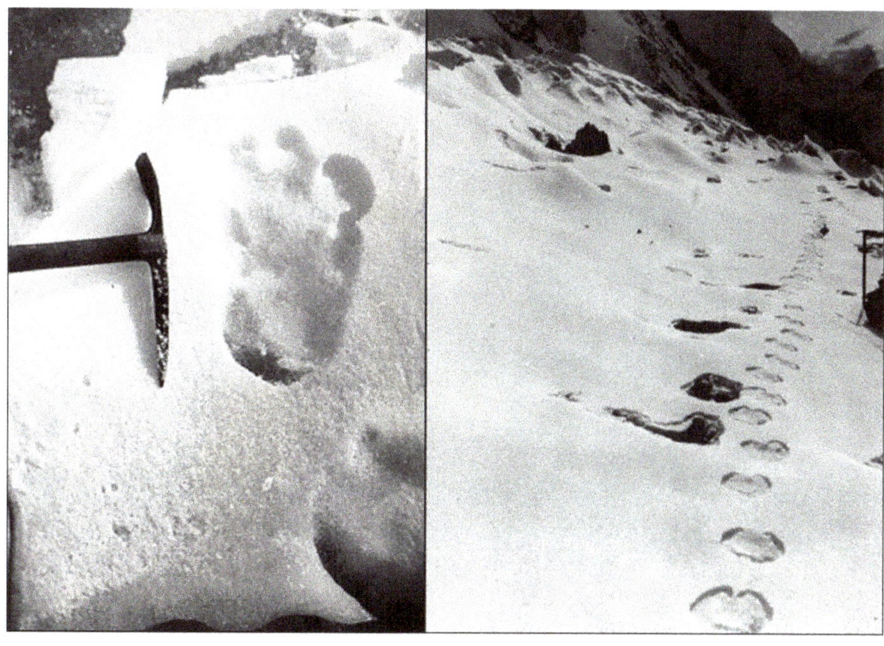

Footprints seen by the 1951 Everest Reconnaissance and photographed by Eric Shipton on the Menlung Glacier.
Chapter 4

A mountainous area of Western Tajikistan.
Chapter 8.

The Tien Shan mountains along the Kazakhstan-Kyrgyzstan border.
Chapter 8.

Kyrgyzstan

CHAPTER 10 • DATA FROM NORTH AMERICA

This chapter serves only to make the reader acquainted with data from North America, where research into Bigfoot has just started.

Ethnographers have written for at long time about the traditions of the Eskimos – that is, the people of the Arctic coast of America from Bering Point to Greenland. They speak about a humanoid creature identical to the yeti, which they encountered when they first settled in these areas. Similar tales are known from Indian tribes in North, Central and South America. Occasionally, not only in the American press of the last century, but even in Russian media, there have been a few reports about hairy wild men. In 1903 a Russian newspaper reported the discovery of "wild men" in America. Local farmers had first noticed their sheep disappearing, and then they discovered the culprit living in a cave in the woods. "He had no clothes on, his body was overgrown with long hair, and he looked like an orangutan."

Despite the reports, there was no attempt to formulate any biological conclusions in America until very recently. It was only in 1958, at the same time as the study began in the Soviet Union, that the prominent zoologist Ivan T. Sanderson began a systematic collection of information in America. Since then the organiser of the Nepalese expedition, Tom Slick, turned his attention to the question of the US Bigfoot in 1960, followed by his collaborator Peter Byrne in 1960-1961. Peter Byrne's preliminary observations are of great interest, but they are only known to me through private letters, which unfortunately I have not had the opportunity to compare with his recent article about the search. As for the material collected by Sanderson, it was originally published as several separate articles, but they have now been organised into a book of several chapters.

This chapter is a series of extracts from the book by Sanderson. These excerpts are by no means exhaustive, but they will give the reader, who does not have access to Sanderson's book, an idea of the type of work he has done and the evidence he has collected.

[Translator's note: What follows in the central part of this chapter is 34 examples of reports from Ivan T. Sanderson's book 'Abominable Snowman – Legend come to life' – but the reports, page references and notes are based on a Russian translation of the book. And since this chapter was written to give a Russian reader an idea of what was going on in America in the late 1950s and early 1960s, I see no real reason to repeat

them. Most readers of this CFZ edition will be either well versed in Sanderson's work, or have easy access to his book. It can even be downloaded in a PDF version from the Internet. The actual translation from the American edition to Russian also seems to be shaky at best, so translating it back to English would make it even more inaccurate.]

After his book was published in August 1961, Sanderson says he received abundant additional information, including new data from South America, which was published quite recently. The American research into relict hominids is still very much in its infancy, but it is developing rapidly, and will probably become an avalanche. It would therefore be premature to try any form of biological discussion or form a conclusion. But there is no doubt that, with more information, the picture will be more complete and clearer.

PART III
Analysis and Synthesis

CHAPTER 11 • PRELIMINARY DESCRIPTION OF *HOMO TROGLODYTES* L. ("BIGFOOT") [50]

(1) Morphology
In Chapters 2 - 5 I presented a general history of Bigfoot. Chapters 6 - 10 provide an overview of the descriptive material separated by geographical areas. Based on this, it is now possible to make a morphological and biological description of the "wild man."

The first who tried to make some form of classification of the many observations and data was Ivan Sanderson. According to Sanderson, the key to scientific systematisation of all the available material is to compare the various creatures to humans, i.e. seeing them as a biological phenomenon. According to Sanderson, the various ABSM (abominable snowmen) can be divided into four groups.

The four groups of Sanderson are as follows:

1) "sub-humans" (Eastern Eurasia: Malaya, Indochina, South China, Central Asia)
2) "proto-pygmies" (Indonesia, India, Africa, possibly Central America and north-western South America)
3) "neo-giants" (Indochina, Eastern Eurasia, North and South America)
4) "sub-hominids" (South Central Eurasia, Nan Shan, the Himalayas, the Karakoram)

For each of these four groups Sanderson gives a list of characteristic features, as well as a list of several types known from the descriptive material.

As a result of the expedition in 1958, Tom Slick concluded, based on data collected by the Byrne brothers, that there are at least two types of ape-man-like species in the Himalayas. One type is the yeti with tracks that are about 30 cm in length, and black hair up to 20 cm in length. The other type is smaller, has shorter hair and a reddish tint. According to Bernard Heuvelmans, assuming the existence of two or even three different types explains all contradictions in the collected material. "While some witnesses claim the snow man is a giant, all the Sherpas who have seen him say he is the size of a human or even smaller." All the information says that yeti – the giant – comes from the highest parts of the Himalayas, close to the border of Tibet. The other, the smaller one, lives in the lower valleys. In 1957 the senior lama Punyabayra, who lives in the city of Kathmandu, said that there are three types of yeti. *Nyalmo* grows to 4-5 metres and lives at the highest altitudes. They are carnivores. *Rimi*, that are no taller than 2.5 metres, live further down and are omnivores. Finally there are *rakshibompo*. They do not exceed 1.7 metres.

Heuvelmans provides some evidence to support the existence of at least the two last types. He is convinced there are two types of yeti that differ in size and colour. In all probability it is variation within one species. Or it could simply be a matter of sexual dimorphism, just as male gorillas are considerably larger than females but can only walk on the ground, whereas the females can climb trees. Thus, the male is the most powerful and has the most massive body, and is better equipped to endure the cold. Maybe only the males dare penetrate the upper zones of the mountains in search of prey. It is possible that the smallest reddish individuals are in fact the young of large black individuals. After all, the colour of gibbons can completely change with age, from white to very black.

I see no need to refute in detail the theory behind Sanderson's four groups of ABSM. They are artificial and will wither away by themselves as we continue our work. The deeper cause of the classification is the general idea among many zoologists that there is knowledge in classification. In my opinion this is not the case. I do not deny in any way that sometime in the future we will be able to identify species or types of local races in the animals we are studying, but today it is a premature assumption. We simply do not have enough material for a justifiable classification. In my view any hypothetical classification is not a help, but will actually hinder our research, and weaken our arguments against those who deny the existence of any form of relict hominid.

I propose a different method, a different way of thinking. Let us study these creatures as a single entity, and imagine a certain variance in many features within a given species. In modern zoology, the study of variations within a single species is still a contentious issue. Some zoologists regard noticeable deviations as sufficient basis for allocation of a new species, while others consider the whole issue a more flexible concept.

We find an excellent example in favour of the latter view in paleoanthropology: the study of "Neanderthal man" (*Homo neanderthalensis*). The studied specimens display huge variations in many different aspects. This could of course be a case of multiple local forms, but after discoveries of fossil remains in caves in Palestine, it is clear that it is also a case of huge individual variation. Why not imagine the same for the "snowman"? With their capacity for travel it is not likely they would develop many separate forms or species.

So the most likely assumptions would probably be:

- If we recognise all the descriptive material in the previous chapters as authentic, all of it relates to the same species.
- There is one highly variable species.
- The variations are probably caused by local variations, age, seasonal variation and sexual dimorphism.

So if it is all one form, to what family should we attribute it?

To do this, we will focus on two groups of data: descriptions and depictions, especially photos and sketches * and the mummified hand from Pangboche.

Scalps

As for the scalps from the Khumjung monastery attributed to the yeti, it is hardly worth retelling the long story about them here. The discovery of them in 1954 seemed a very important event in the study of the Bigfoot problem. There were photos published, data on the size and the peculiarities of skin and hair. Some experts, such as Professor Wood Jones, argued that the scalps could be artificially made from the leg of any large ungulate. Heuvelmans displayed a lot of wit and inventiveness in his attempt to substantiate the authenticity of the scalps.

It is well known that the particular branch of comparative anatomy which studies hair has not developed a taxonomic method. You cannot say from a hair sample whether it belongs to a predator, rodent, ungulate or primate **. Experts have to compare hair samples with samples from different animals. Heuvelmans suggested that the animal from which the leather for the scalp was made was in fact a rare species of mountain goat, *Capricornis sumatraensis*.[51] A commission of experts gathered to study the scalp from Khumjung that Sir Edmund Hillary brought to America and Europe came to the same conclusion. It is quite possible that rivalry between the various monasteries had encouraged the making of an artificial product to attract people and pilgrims. And that means that somewhere there may be originals.

The conical shape of the scalps has played a significant, and not altogether positive, role in the various reconstructions of Bigfoot. English and Czech anthropologists have tried to reconstruct the shape of the skull and the head on the assumption that the scalps were original. Tschernezky went even further, and suggested the outline of the body and the limbs. It was extremely awkward. Unfortunately these reconstructions became widely known, and influenced the imagination of those who interpreted the Bigfoot as a monstrous anthropoid quite unlike anything living today.

Even though the scalps as such do not tell us anything about the shape of the head of Bigfoot, the shape of them is no accident. We have descriptive material from the Himalayas and elsewhere that repeatedly describes the conical head of these creatures. It is unlikely that this can be explained by the specific form of the skull; the answer must lie in some form of deformation of the skin. There might be a sagittal crest on top of the skull, which may contribute to the shape, but also a roll of subcutaneous fat. This may be a novel idea, but it is possible. I base this on the

**Translator's note: There are no sketches or photographs, or indeed, figures and tables, in the manuscript. Presumably the author never got around to making and inserting them.*

** *Translator's note: This is no longer true.*

description of a very peculiar sleep posture in a Bigfoot. This comes from V.A. Khakhlov, and describes a female his informant was able to watch almost every day for several months. When it was sleeping (presumably in the afternoon), it would rest on its knees and elbows, and the top of its head, while its hands would be folded around its neck. There is some support to this story in the fact that this sleeping posture is sometimes seen in very small children. Extra padding on the head, as well as hardened skin on the knees and elbows, would of course facilitate this position. And Khakhlov has in fact described the skin on the knees and elbows of the Bigfoot as similar to the soles of a camel.

Foot, locomotion

Let us now turn to the extensive series of photographs of the tracks of this creature (mostly found in snow), casts, sketches and measurement and descriptions of the tracks.

The tracks of Bigfoot have been a matter of debate for quite some time, and most of it has been wasted. But thorough studies, by this author and others, have shown that the bulk of the Bigfoot tracks cannot be attributed to bears or the large ape of the Himalayas, the langur.

The validity of the "classic" pictures taken by Eric Shipton in 1951 has been confirmed by abundant control material, although none of it is of the same quality of the Shipton photos. An interesting addition is the tracing of a plaster cast brought back by the 1958 Tom Slick expedition, and the North American material of snapshots and sketches of tracks belonging to Ivan Sanderson and Peter Byrne. There is some preliminary material from the Caucasus as well. In general, it is a significant series of tracks, but found and documented under very different circumstances.

The main question is whether it is possible to attribute all these tracks to members of the same species or type of living creatures, or whether they exhibit such fundamental differences that we have to assign them to substantially different types. Ivan Sanderson for instance agrees with "Russian scientists", that almas tracks differ little from those of Neanderthal man, and he therefore regards the almas as being descendants of Neanderthals. Tracks of similar type can be found in America and Africa. They look very much like those of modern man who has never worn shoes. Only a very detailed analysis can reveal the differences. As for the Shipton track, Sanderson sees this as clearly different from a human, modern or Neanderthal. So this has got to be another species. But Sanderson has to a large degree based his ideas on the investigations of V.Tschernezky, and therefore cannot be considered objective. If we truly want to understand this, we have to know the history of the study of this subject.

In his first article from 1954, Tschernezky made an important comparison between the contour of the foot of the Bigfoot from the Shipton photo and the track of a gorilla, where he found significant differences, and also with fossilised Neanderthal footprints preserved in the "Cave of the Witches" in Liguria and published by Professor A.K. Blanc in 1952.[52] In spite of significant differences in the absolute size of the individuals, the tracks, according to Tschernezky, nevertheless "show the greatest similarity."

Despite this result, Tschernezky still concludes that there is no close relationship between Bigfoot and hominids, and that it therefore has to be assigned a particular genus and family, while there is no doubt about the close connection between modern humans and Neanderthals. The conclusion is based on the combination of human and monkey traits in the tracks. Monkey traits: the big toe is very short and deflected to one side. Human traits: short toes and a broad general outline of the foot. Among characteristic traits in Bigfoot tracks is an extremely broad and solid heel, and the ratio between the length of the foot and its width at the toes.

A second article by Tschernezky from 1960 is specifically devoted to a reconstruction of the foot of the Bigfoot based on the Eric Shipton photography. Based on the outline of the tracks, Tschernezky made a plaster model of the foot, used this to make tracks in the snow, and compared these to the Shipton photos. They were very similar to the natural tracks. This idea is basically flawed, as you cannot compare the tracks of something made with a hard model to a soft and movable natural foot. The similarity of the artificial tracks and the natural ones does not prove that the mechanics of their formation was identical.

This basic mistake has since misled Ivan Sanderson and several others to publish this unnatural foot as a "true" reconstruction.

We can however note the following. The front part of the foot is very wide (about 43% of the length). This relationship, as rightly pointed out by Tschernezky, can also be found on the foot of the Neanderthal from the Kiik-Koba cave in the Crimea. The metatarsals are short compared to a modern human foot, and the phalanges are longer. The depth of the indentation made in the snow by the big toe suggests that it bears a significant part of the body weight. However, it may also be used to grab objects or aid in climbing. The second toe is longer than the first. In modern humans that is only seen as a deviation from the norm. It is usually associated with an increased ability to grip things. An upright walking Bigfoot is probably unstable, and is likely characterised by a greater use of the elongated second toe in balancing than in modern humans.

Tschernezky correlates the weakness of the imprint of the fifth toe with the fact that, in the modern human foot, the two peripheral phalanges of the fifth toe are often fused together. Based on forensic evidence of tracks of a bare human foot, Tschernezky also states that where the back of the heel touches the snow it often casts up a little mound of snow. The clearest imprint of a human footprint is the outer side of the back of the heel, and the inside of the big toe. The weakest part is the outside of the foot near the little toe, and the inner side of the big toe. These details are clearly visible on tracks and plaster casts of Bigfoot tracks. Thus the general data indicates that the Shipton Bigfoot in general walks like a person, although a comparative anatomical analysis points to special features of its feet compared to ours.

So, if we ignore the above-mentioned errors in the reconstruction, Tschernezky's main observations lead to only one possible conclusion: the track left by the Shipton animal is clearly hominid, although it varies in different ways from *Homo sapiens.* But, rather strangely, Tschernezky ends his article by concluding that Bigfoot is probably similar to the fossil Gigantopithecus.

It may be that Sanderson and Tschernezky, in the analysis of the yeti tracks, pushed it too far from the human race, but we have to admit that the anatomical analysis of the plaster casts of the footprints from British Columbia and Northern California is quite convincing, and they look very human at first glance. But they are special. They do not have the toes spread out in a fan-like configuration like the footprints of a bear, but neither are they spread out from the line of walk as in humans. Instead the toes are pointing directly ahead in the direction of the line of walk. Further analysis shows that although the tracks look fairly long, they are in fact relatively short and wide (length to width ratio about 1.6). Two pads can clearly be discerned behind the toes. This is in itself rather unusual. So is the relative lengths of the toes, as they all reach a line drawn from the tip of the first toe to the tip of the fifth. This is not something we see in humans. There is also no sign of any dirt being squeezed up between the toes, something that is extremely common in human prints. Sanderson tries to explain this by suggesting the existence of some form of membrane between the toes. Or it may be explained by the creature pressing its toes together with greater force than in a man.

A few mistakes crept into Tschernezky's analysis, probably because he only took one of Shipton's footprint photos, and did not use any other images or descriptions of tracks of these creatures, and he only used one cast of one foot of the Ligurian Neanderthal isolated from all available material of the feet of the Neanderthals.

If we compare the Shipton photos, we see some common features. 1. The absence of an imprint of toe 5. This may be associated with increased extension of the big toe – especially in deep snow. 2. The rotation of toe 1 towards the inside of the foot. In the cast, the big toe is tightly pressed against the second as in the human foot. Tom Slick and Peter Byrne noted the "Y" shape, which resembles a fossilised Neanderthal track from Liguria. Geologists and hunters have described similar tracks from the Pamirs. This bulging or "reclined" toe is different from bear or human tracks. In the various tracks we also see signs of a significant force pushing from the side of the big toe.

Thus it is a major error that Tschernezky and Sanderson based their analysis on a single photo and a single track. They did not take into account the differences caused by different walking speeds on different types of ground, nor factors like different inclines and softness of the surface. The same two-legged being could leave a large range of different tracks at different times or in different conditions.

It is quite understandable that there have been difficulties in analysing tracks of relict hominids. Although the science of ichnology, the scientific studies of tracks, is well-established, as it is of importance to forensic scientists, hunters, naturalists and paleontologists, it is based on studying and following the tracks of well-known, not unknown beings. Reconstructing an animal based solely on tracks is not something that is done on a regular basis. Only paleontologists have been facing these kinds of difficulties, and will be the first to acknowledge that they need to know more.

Based on a study of photographs of series of tracks, sketches and measurements, we can try to answer the question regarding the movement of the big toe and the foot of this creature in relation to the foot of the Neanderthals. The exemplary work of G.A. Bonch-Osmolovsky and B.B. Bunak has shown, on one hand, that the lateral mobility of the big toe of Neanderthals and other paleoanthropic species was not as great as the flexibility and grasping power of anthropoids, but on the other hand was greater than found in modern humans.

We shall now look at two other points: the relative length of the big toe, and the placing of the toes in relation to each other. As we have seen, Tschernezky drew attention to the fact that the big toe of Bigfoot is relatively short, and attributed it to its "simian features". But is this a sign of monkey features or of a feature of early humans? It turns out we see the relatively short big toe in many cases, and it has been noted by many eyewitnesses. We also have at our disposal two tracks of a "wild man" in the Caucasus, both made in the summer of 1960 on wet ground. They belong to two individuals of different sizes. One is found on a mountain, the other in a valley. Both of them have a big toe that is pushed forward. The other toes cannot be seen. They are probably raised above the ground.

We also see a large toe in paleoanthropic tracks. This can also be seen in the images and descriptions of relict hominid tracks. Khakhlov: "Big toe noticeably bulkier than the rest". Zhamtsarano: "big toe unnaturally thick". Shaimkulova: "The first toe was bigger than a man's". Shalimov, the geologist: "Big toe much larger than the rest." This, however, coincides with the characteristics of Neanderthal bones. The bones of the other four toes are also generally greater than in humans. "The toes are flattened compared to those of modern humans and anthropomorphic monkeys. They are characterised by a marked flattening of the end of the terminal phalanges." The flat section undoubtedly serves as a support for large nails.

There is a very clear parallel between the placing and the overall mobility of the toes in Neanderthals as well as in Bigfoot. "The Neanderthal has a greater scope of articulation in the metatarsal-phalanx of the back foot, and in the lateral mobility of the foot compared to modern man. The inter-phalangeal back-planter and lateral

movement were less restricted than in modern humans, especially in the junction of the side and middle of the phalanges, with total mobility increasing from toe 2 to toe 5. The mobility was basically different from that of modern man, especially in the increased mobility of the lateral toes." G.A. Bonch-Osmolovsky rightly wrote that Neanderthal man, compared to modern man, was characterised by "a powerful outstretched foot with free lateral movements. Many of the structural features indicate that toes 2 and 4 differed markedly from those of modern humans and anthropomorphic apes". Here we see a parallel in the tracks and descriptions of tracks from the Bigfoot. "The toes were widely spaced in the tracks," Mongolian scientists reported after a visit to their camp by hairy naked human-like creatures. One of the main differences of the foot of the *ksy-gyik* compared to humans, according to the Kazakhs interviewed by Khakhlov, was the widely spaced toes. According to Pierre Bordeaux, who has analysed Bigfoot tracks, "the toes do not completely merge during walking." Based on sketches and observations of the track of the *kaptar*, V.K. Leontiev writes: "The four toes of the foot are not adjacent to one another as in humans, but have moved apart. The distance between them varies from 0.5 to 1 cm."
Another interesting relation is the width of the foot compared to the height of its arch. Several authors have pointed to the fact that the feet seem exceptionally wide, and the arch very low or non-existent. W. Tokhta: "the tracks looked like humans, but they had flat feet…".

To sum up the question of relict hominid tracks. If you take the entire amount of data available, in 85-90% of the cases there are no pictures, only verbal descriptions, and they do not give any kind of detailed information, only the most common and basic. But in general they give the impression the tracks look like human tracks. The most obvious is the large size of the first toe. This also excludes confusion with tracks of bears. It is also in marked contrast to the tracks of even the largest humans. Although we still have to allow for very large intraspecific variations in size and age of the various individuals. Only 10-15% of the material provides additional information. Closer study has shown marked differences between the relict hominid tracks and modern humans, and marked similarities between relict hominid tracks and Neanderthal tracks. However, this is not to be understood as proving that Bigfoot and Neanderthal man are one and the same. It is quite possible that the same differences and similarities are typical of other fossil hominids and pre-hominids for which we do not have the same amount of fossilised material or tracks.

The Pangboche hand *

- Phalanx (pl. phalanges): The individual bones in the fingers. There are two in the thumb and three in the other fingers.
- Metacarpals: The bones of the hand between the wrist and the fingers.
- Proximal phalanges: Those closest to the hand, joining with the metacarpals.
- Distal phalanges: Those at the end of the finger.

* *Editor's note. Again, I think it worth adding a few definitions here to explain some of the technical terms. The thumb counts here as finger 1. The reader should be warned that the Pangboche hand may well be human.. DNA: further testing by Prof. Bryan Sykes showed it was Byrne's own DNA. The identity of the hand remains open.*

- Condyle: Each phalanx terminates at the distal end in two knuckles called condyles, with a groove between them.
- Epiphysis: A distal phalanx ends in a rough cap of bone call the epiphysis.
- Radial-palmar: the radial side of the palm i.e. towards the thumb.

We now move on to another type of material, namely the mummified hand from the Pangboche Monastery in Nepal.

In 1954 the English Tibetologist Professor Snellgrove first heard about the mummified hand from the monks in Pangboche. He was told it was larger than a human, was kept in the monastery wrapped in several layers of cloth tied with a cord, and that it would be an act of sacrilege to unpack it, as it was a sacred relic. In 1958 and 1959 Peter Byrne was allowed to see the mummified remains being unpacked, and to take several photographs. It turned out to be impossible to get an X-ray unit to the mountain monastery, but members of the expedition managed a partial dissection on site. They removed several samples of dried soft tissue, which was taken to the United States for a microscopic and serological study. In 1959 we were capable of performing an independent parallel morphological analysis based on photos and the preliminary results of the U.S. laboratory.

Before presenting the histological and anatomical results, we have to look at the further story of the Pangboche hand. I have only recently learned from B. Heuvelmans, that Peter Byrne has confessed to a great transgression to science. In 1959, when he was allowed to photograph the hand, he had brought with him a set of bones from a normal human hand. Apparently he had inserted several of the normal human bones into the hand to "reconstruct" it, after having removed some of the original bones. He had not only lengthened some of the missing bones, but had completely replaced the phalanges on fingers 1 and 2. This was apparently done to allow anatomist and primatologist Dr. Osman Hill in London to study the authentic phalanges. However, whether by mistake or by design, Dr. Osman Hill did not just receive true Pangboche bones. This was established when Dr. Hill kindly sent photos of the bones to a Soviet anatomist. They did not deviate in any way from modern man, as some of the bones of the Pangboche hand can clearly be seen to do in the photographs. It is difficult to explain this substitution of the bones, and forces us to be very careful when analysing the newer photos of the Pangboche hand in its "restored" state.[53]

The new photos were taken in 1960 by professor of anatomy at the University of Tokyo Teizo Ogawa, who headed a Japanese expedition in Nepal to study the problem of the yeti. As he did not know about the previous photos taken in the same Pangboche monastery in 1958 and 1959, Prof. Ogawa believed it to be the original hand. At the request of Soviet scientists, he has kindly sent his photos for analysis. Whereas Peter Byrne's photos are very difficult to analyse, Ogawa's photos allow us to compare the lengths of the various bones.

[Translator's note: Porshnev here attempts to distinguish the genuine parts of the hand from those substituted by Byrne. The result is confusing and contradictory. In the original manuscript there are several empty spaces, which suggest to me that Porshnev was planning to insert various other comments, and that he knew full well this was not clear. It has therefore been omitted.]

Another series of photos, which can be called the third series, was filmed in 1961 by an expedition led by Sir Edmund Hillary. These photos have not yet been made available to me, but there is a drawing made from one of them in an article published by Professor Vallois. This drawing indicates that the hand was in the same condition in 1961 as in 1960. It is valuable because this is the first photograph showing the restored hand with the palm side up, although the caption says it is "the backside of the left hand".

The preliminary results of the X-ray diffraction studies suggest that the hand is more than 300 years old. The cellular structure is severely damaged, and it is difficult to see any details with the help of X-rays. The serological studies, led by Professor Charles Leon at the University of Kansas, did not give any definitive results when comparing the hand with material from human, gorilla, and several Old World monkeys, as well as cattle, pig, goat, horse and bear.

Regarding the morphology of the hand, the preliminary findings of the foreign experts tend to point in one direction, although there are some contradictions. Unfortunately not enough details have been published to allow a firm conclusion. All agree that the basic features of the hand are hominid, but not that it belongs to a human of modern physical type. Everyone agrees that it is "almost" human. Dr. Osman Hill of the London Zoo defines it as human, that is not monkey, assuming that it can show sub-human traits. In his last article from 1961 Dr. Osman Hill gives this rather contradictory statement: "The mummified hand which is kept in the Pangboche monastery proved human, though its metacarpal bone in the available photo shows a number of features similar to anthropoid apes." However it should be noted that, in a letter to me dated 30 July 1962, Dr. Sanderson reports that the personal judgement of Dr. Osman Hill of the Pangboche hand is that it is human, but so aberrant that it is most similar to the hands found in the Crimean Neanderthal Cave (Kiik-Koba). This demonstrates the ongoing work and the problems caused by an attempt to sneak in fake material. Dr. Hill's final statement is of course of great satisfaction to me, as his conclusion is so close to mine.

The anthropologist Dr. G. Agogino is inclined to conclude the hand belongs to an unknown primate, and that if it is human it must have belonged to a very strange abnormal man. He cites the following features. The hand is wide compared to the overall length. Metacarpals are too flat to be human, and massive as in a gorilla. Condyles have a deep depression and are larger than in even the biggest humans. Metacarpal and phalangeal proportions are slightly different from normal human proportions. Heuvelmans considers these features sufficient to distinguish the hand from that of a modern man, and that based on the photos it looks very close to the hands of the Neanderthal from Chapelle.

So, what conclusions can be drawn based on the anatomical studies? First of all, a number of important features of the skeletal hand clearly points to it being a hominid. These include the roundness of the heads of the metacarpals, the general shape of the bones, the clearly expressed radial pattern, the relatively large size of finger 1, and several minor differences. But at the same time some features clearly deviate beyond the variation found in modern man. The more primitive or monkey-like features of the Pangboche hand include the following:

- The proximal phalanges on fingers 2 and 3 are elongated compared to the metacarpals, so that the metacarpal/phalanx index, if the photos do not give a false impression, is well outside the range of humans.
- The strongly marked groove in the sagittal plane on the head of the phalanx in finger 2 is clearly different from the corresponding area in a human.
- The base of metacarpal 2 is markedly narrower than it should be on the hand of a man, compared to the very broad bases of metacarpals 3 and 4.

In addition some features have also been noted, but they could be explained by the drying and wrinkling of the mummified soft tissue:

- The weak development of finger 1, especially the distal phalanx.
- The restriction gaps.
- The elongated nails on fingers 1 and 2.

There is no doubt that the hand is broad and massive compared to a normal person, and there is no indication of acromegaly. In general there is nothing to suggest anything pathological. Judging from the photos the first finger also seems remarkably little opposed to the others. These initial anatomical findings clearly point to the fact that the Pangboche hand is somewhat primitive, and not from a modern man. Nor is it from any of the great apes.

To answer the question of the systematic position of the creature, we must also turn to the examination of the scalp. The study of the hair samples could not relate them to the hair of any modern ape, whereas they did show some affinity to humans. The same structure has also been found in several other samples.

On the basis of the examinations and the 1958-1959 photos and studies, we should be able to form at least one undeniable conclusion: that not more than 300-400 years ago, there was at least one individual of this species living. It clearly belongs to the family Hominidae, but is clearly more primitive, seemingly belonging to the Neanderthals broadly speaking. According to the lamas this yeti still occurs today, and we have no reason to reject this supposition.

Next we must look at the photos of the Pangboche hand taken in 1960 by Teizo Ogawa.

These photos are of a considerably better quality than the first ones. One can even see the soft tissue preserved on the palm side. As in previous photos the third finger is covered with dry skin. There are some differences when compared to the first set of photos, but they are probably due to the "restoration" attempted by Peter Byrne. Perhaps the most important difference from the previous photos is the fact that the entire first finger (both metacarpal and phalanx) is no longer covered by mummified remains. This opens up a great new opportunity for study.

Again, at the first look on the new pictures the massiveness of the metacarpal bones is immediately visible. It is even clearer than in the previous pictures. This makes it abundantly clear that this hand does not belong to a modern man, but to some ancestral primitive form. When reading descriptions of Neanderthal skeletal material, all authors have emphasised this feature of the metacarpal bones: they were "massive and stocky" – more so than the average *Homo sapiens.*

There are important features to be found on the first metacarpal. Among them is the massive distal end and what looks like extra bone growth on the radial-palmar side. This bony prominence is undoubtedly an attachment point for particularly strong muscles.

[Translator's note: Here follows a series of short notes about the possible evolution of these muscle attachments, and whether they can tell us anything about the evolutionary distance between the owner of the Pangboche hand, the Neanderthals and modern humans. The notes were probably meant to be worked up to a detailed discussion later on, so I have omitted them here, although there is a preliminary conclusion, which is that the muscle attachments of the Pangboche hand seem more primitive/archaic than Neanderthals.

In this section of the book, we find a number of references to photographs, drawing and graphs, but none of them is actually in the book. Presumably they were meant to be put in later.]

The strong development of the muscles of the thumb, as evidenced by the large curvature in the metacarpal bone to the palmar side (the same applies to the proximal phalanx of the thumb), can also be seen in the first metacarpal bone from the remains in Kiik-Koba. Bonch-Osmolovsky wrote that the curvature is greater than in all others, followed by another Neanderthal, gorilla, modern man, chimpanzee and orangutan. The curvature of the bone in the Pangboche hand is closer to what is found in Neanderthal hands.

We will now turn to the most important feature of the first metacarpal bone in the Pangboche hand. This is perhaps most diagnostic when compared to fossils of Neanderthal hands. All the experts who have studied Neanderthal hands (Boole, Sarazin, Bonch-Osmolovsky) have remarked on a strange feature seemingly characteristic of at least three hands: one from Ferrassie, one from Chapelle and one from Kiik-Koba. In all Neanderthal hands, when comparing the first metacarpal bone with any of the other bones in the limb, it is slightly shorter than in modern man. It is a small but undeniable fact. Based on a comparison of data from his doctoral dissertation "The skeleton of the hand of primates and man" (Leningrad 1950), and measurements made by him on the Pangboche hand, L.P. Astanin came to the conclusion that the proportion of this bone is of great theoretical interest for anthropology. Measurements on the pictures have shown that the first metacarpal bone is shortened:

Index 1: length of metacarpal 1 as % of the length of metacarpal 2

Modern man	69.8
Pangboche	62.6
Gorilla	55.2
Orangutan	48.6
Chimpanzee	47

Astanin concludes that the Pangboche hand is just in the middle between man and gorilla. Some researchers have reservations regarding the measurements, as they think it is not possible to discern the ends of the metacarpal bone clearly. Others have no problems with the measurements.

Astanin emphasises that the reasons for the shortness of the first metacarpal bone in a gorilla and in the Pangboche hand are quite different. In a gorilla the metacarpal bone 1 is shortened because the whole first finger is somewhat reduced, and the finger is shortened even more than the metacarpal bone. As for Pangboche, the metacarpal bone is shortened not only in relation to the metacarpal bone 2, but also in relation to the length of finger 1:

Index 2: length of finger 1 as % of length of metacarpal 1

Modern man	110
Pangboche	120
Gorilla	96

The entire first finger of the Pangboche hand is massive. The main phalanx of the thumb is not typical for modern man with the expanded distal epiphysis (head). Note that this feature was also observed on the photographs taken in 1958: the main phalanx of the thumb seems unnatural. The shape of the end of finger 1 clearly deviates in shape from a human. It is markedly extended and spatulate in the distal end. The latter, of course, serves as a support for a very large coarse nail. It closely resembles the amazing features of the terminal phalanx of the Neanderthal from Kiik-Koba on which Bonch-Osmolovsky suggests a huge nail has been growing. Prof. Astanin suggests that the massive and short terminal phalanges on finger 1 are a tool for digging roots: "It is clear that it is easier to dig in the earth with the short and thick first finger compared to the long and thin second finger."

It is impossible to discuss finger 2 of the Pangboche hand, as the terminal phalanx may not be the authentic one; however, this does not radically affect the very interesting results L.P. Astanin got when he measured it. It was found that the relative length of finger 2 is higher than is found in modern humans and apes:

Index 3: length of finger 2 as % of the length of metacarpal 2

Chimpanzees	116
Gorilla	117
Modern man	118
Orangutan	124
Gibbon	133
Pangboche	145

One indirect circumstance does suggest that finger 2 on the Pangboche hand is unusually long. It is the relative length compared to finger 3. Finger 3 is certainly genuine, but is also bent, which makes the actual comparison very difficult; but we applied various techniques to help us, using not only compasses and thread, but also by reproducing the same position of finger 2 and 3 in various persons. It turned out that, in contrast to modern man (and other primates), finger 3 on the Pangboche hand is not longer than finger 2. But as to what this indicates, we do not know.

As we have seen, a significant number of features distinguish the Pangboche hand from the hand of a modern man. It is in many ways closer to a Neanderthal hand. But an important question still remains – is it in fact within the variations of the hand of a man? Some of the measurements are within range, but not everything put together. Even if we assume that every single feature of the Pangboche hand is within the variation range for modern man, the combination of features found in the Pangboche hand is not found in modern humans.

The Pangboche hand is different from the hand of a modern man to the same extent a Neanderthal hand is. Most likely it should actually be classified precisely as Neanderthal or Neanderthaloid. I even allow myself to think that the Pangboche hand is able to shed further light on the general discussion regarding the anatomy and development of the Neanderthal hand. Prof. Astanin has for instance suggested that the Pangboche hand lacks any kind of specialisation.

It could be argued that it is impossible to make any generalisations of the anatomy, as we in fact only have access to a series of pictures, and there is no hope of the bones getting into the hands of anthropologists in the near future. This limits what we can actually study. We do not see all sides of each bone. We cannot make any comparisons in relation to the absolute size of the bones, as we do not have any reliable scale which could help us to determine with precision the actual size of the Pangboche hand. We only know it is a little smaller than an average human female hand. But there is plenty of opportunity to measure the proportions of the bones, that is the relative values as well as the anatomical configuration, their relative position etc.

The main thing to realise after studying the photos is that we are only dealing with photos and not the actual specimen. Our analysis of the Pangboche hand cannot be confirmed, and we must also bear in mind that those who take the pictures often take them expecting to find support for hypotheses of an unknown "terrible anthropoid", rather than something of a Neanderthaloid nature. But we know from other cases that it is quite possible to make accurate observations and conclusions based on measuring bones in photographs when the original material or casts have not been available.

[Translator's note: There are several sections in this part of the book, where the author discusses various anthropologists ideas regarding the possible level of intelligence and consciousness in Neanderthals, but every time concludes that this has absolutely nothing to do with a purely anatomical discussion, which is after all what this section is all about, so there is really no need to discuss it. One gets the impression this was meant to be deleted in the finished edition of the book. I have therefore decided to delete it here. Secondly there is a rather lengthy discussion of the anatomy of the hand of Neanderthals. As I have stated before, there is really no need to repeat this, as our current knowledge of Neanderthal anatomy is far greater than it was in the early 1960s. We no longer consider Neanderthals as almost ape-like, which they clearly did at the time. I have therefore omitted that as well.]

We can now try and compare the anatomy of the Pangboche hand with modern descriptive data. We have already mentioned a large number of references in folk legends to the custom of cutting off the hands of killed "wild men" as trophies and talismans.

In June 1948, the geologist A.P. Agafonov was working in the area around Sary-

Chelek Lake (Tien Shan, Chatkal Range). Here an old blind shepherd named Madyer showed him an ancestral heirloom, a dried severed hand which Madyer's grandfather had cut off a huge red monkey. According to Agafonov, the general structure of the hand was human, but the back side was covered in dense brown hairs up to 1 cm in length. Mattuk Obderaim described to Prof. B.A. Fedorovich how he had seen a *jawo-halga* (wild man) killed by his uncle in 1944. Among other things he had noted that "the thumb was closer to the other fingers than in a person." A shepherd called Mahoma reported on 13 August 1959 on female *kaptars* he had seen passing through the main Caucasian range between Dagestan and Azerbaijan. About the hands he said that there was hair growing on the back until the start of the middle finger. The fingers were longer than in the average person and closer together.

When V.A. Khakhlov recorded the stories of the Kazakhs, one informant described the hands of the wild man as very narrow compared to the length of the fingers. "To show what it looked like, the informant compressed his hand at the base of the fingers and turned the thumb in towards the palm. This made his hand look long and narrow and somewhat resembling the hand of a monkey." From these stories Khakhlov got the impression that the palm of the "wild man" looked like an elongated rectangle. Based on Khakhlov's information fingers 3 and 4 are the longest and of the same length, whereas fingers 2 and 5 have the same length. "The thumb is quite long and, apparently, lies almost in the same plane as the palm. In any case, it cannot be opposed and give a true grasping motion, as in a man."

This morphological description of the added functionality is a very important observation. Judging from stories from various witnesses, when the wild man climbs he holds his hands like a hook. In this way he is able to scramble easily over bushes and trees, Further evidence comes from the Kazakh who saw the wild female on a leash.[54] When she grabbed an object or insect, she pressed the thumb towards the side of the middle phalanx of the index finger, and did not press their ends together as a human would. Based on this information Khakhlov believes that the "wild man" is a very strong climber.

This is to a certain extent contradicted by stories of people escaping from a Bigfoot by climbing a tree. A creature can easily cling to branches in a tree, but climbing up a fairly smooth trunk is next to impossible without the use of an opposable thumb.

The description of the "wild man" hand as given by Khakhlov would be incomplete if we did not mention the nails, which he describes as "narrow, long and prominent." In all other descriptive material we repeatedly hear the words "strong", "large" regarding the fingers, as well as indication of the nails being exceptionally long compared to human nails. Finally, there is the question of hairiness. Every single testimony repeats the fact that the palm of the hand has no hair. Regarding the back of the hands, the information varies, some just refers to the hands as being hairy, others quantify it, such as only up to the first joint of the fingers and so on.

Based on the various data we conclude that the Pangboche hand is close to the hands of paleoanthropic hominids.

The body and the head
It would highly desirable if we could supplement our data on the Bigfoot with measurements and studies of the skull. It is after all fundamental for the science of taxonomy in zoology as well as in anthropology. Naturally researchers have been looking for ways to establish the morphological features of its skull.

In the beginning the research went in the wrong direction. Experiments trying to reconstruct the shape of the skull based on the Bigfoot scalps did not give a positive result. The remains are controversial, and some researchers do not consider them to be real remains.

Some attempts have been made to make a forensic facial reconstruction with a technique developed by Professor M.M. Gerasimov [55], and based on a Buddhist ritual mask found in a zoologist's collection, which is believed to be a true representation of a Bigfoot skull or face, possibly made by some form of papier-mache layering. However, most anthropologists and primatologists have strong objections to this approach.

[Translator's note: A series of short notes suggesting the use of Neanderthal material as a starting point for studying and reconstructing skull morphology follows in the original manuscripts, but has been deleted here.]

However, we do have some solid points of reference for a description of the morphological characteristics of the relict hominid and a determination of its taxonomic position. Our conclusion can be summarised as follows. Morphologically we are dealing with a creature no further from modern man than the australopithecines, but not as closely related as the Neanderthals.

There are several distinguishing features we can extract from the immense mass of oral communications about the Bigfoot and compare with modern man on the one hand, and on the other hand with any of the monkeys.

In all the descriptions, the following characteristics distinguish the creature from modern man:

- hairiness of the body (easily observed due to lack of clothing)
- lack of articulate speech

In relation to monkeys and apes, the Bigfoot is distinguished by:

- Bipedalism

- the presence of developed breasts in females (generally absent in all monkeys, although in exceptional cases breasts can be developed, even to such an extent that they will be sagging, but only during lactation).

Bigfoot is distinguished from the lower monkeys by the lack of a tail.

Although the relict hominids have developed bipedal gait, it is somewhat incomplete compared to the gait of modern man. We have already given a number of descriptions of their rather clumsy rolling gait with fairly spread out legs and slightly bent knees.

We must also add the fact that they seem to be rather short-legged.

The general impression of the figure of the Bigfoot is a being with relatively longer arms than normal humans, but this is probably an artefact caused by a more forward slanting body posture. What is considerably more important is the way Bigfoot uses its forelimbs when moving.

The hands are used in locomotion in a variety of situations: when floundering in the snow, jumping over cracks, running away from danger and persecution, climbing on rocks or similar. It is very characteristic, that in these situations there is no tendency for Bigfoot to rest on the back of the knuckles of the hands. The creature always rests on the palms of the hands. In the case described by the geologist M.A. Stronin, the creature ran up a steep slope, helping itself along with the forelegs ("tossing them under the body like a horse galloping"). Other observers have noted: "They used their hands a lot, especially when climbing up the mountains." Judging from the stories, a Bigfoot often takes off running on all fours, even on a flat surface, but the actual use of the forelimbs is most clearly seen during movements over cliffs and steep slopes. In these cases, the lower extremities just support the body, whereas the forelimbs pull it along. A "wild man" seen climbing over a fence in the Caucasus first pulled itself up by its arms, and then threw its legs up.

But despite the many references to the use of the forelimbs during locomotion, it is still an insignificant percentage of the whole. Bipedalism seems clearly to be the main form of posture and locomotion. Bigfoot can climb mountains on their hind legs, walk for many kilometres, run fast and jump up to 1.5 metres. A Bigfoot can sit like a man, mostly on its haunches, but also directly on the buttocks, as well as sliding down slopes in this posture. But to the biologist there is still an important difference in these relict hominids compared to humans: they use their forelegs more and climb more than humans.

The structure of the body of a Bigfoot is quite unique. In most cases, the first impression of an observer is that the figure is clearly humanoid – like a person. But on closer inspection, there is also something of a beast mixed in. The various testimonies often highlight the general massiveness of the body, the powerful physique, and the muscular, stocky and powerful chest region. In a few very rare cases, the eyewitnesses have described individuals as being very thin and skinny or in a state of poor physical development. This is of course understandable, as there will always be individuals deviating from the more conventional physical type as a result of poor nutrition, illnesses etc.

As already noted, there is a particularly large variation in the physical size of Bigfoot. Significantly more than 50% of the entire series of available evidence suggests individuals of the same size as an average person (1.5 – 1.7 metres in height). However, there are also a considerable number of cases describing a size above the average human (1.8 – 2.2 metres) and some supposedly even reaching up to 2.5 – 3.0 metres. In other cases we find indications of individuals of lower stature than humans, but this could easily be explained by young individuals not yet fully grown. The only attempt to judge the weight of these creatures was done by Wyss-Dunant studying tracks in snow. This led to the tentative assessment of the weight being 80–100 kg.

The descriptive data also show a significant sexual dimorphism (i.e. size difference between male and female). The females are often also described as having very pendulous breasts. The descriptions of the younger hominids are too sparse to enable us to determine any special structural features.

Males, females and young relict hominids are covered with hair from head to toe, apart from the soles of the feet and the palms of the hands. When it comes to the finer details, there is a lot of variation. This could of course be a result of the imagination of the storytellers, but Prof. V.A. Khakhlov has formulated a more rational hypothesis. The colour and character of the fur of an animal, wild or domesticated, can change quite dramatically between summer and winter. One can even find traces of these changes in modern humans. So it would be safe to assume that similar changes occur in Bigfoot as well. So we can assume seasonal variation as well as individual variation, variation within a population and with age in the species *Homo troglodytes* L.

More than half of the eyewitnesses have described the hair colour of the relict hominids as red-brown or brown. For the rest, there is a quite significant variation. The colour is often darker, sometimes described as “brown-black” or simply “black”, but also lighter: light brown, yellow, yellow-grey or grey-fawn. Finally a few grey-white, possibly albinistic individuals have also been described.

Based on the various descriptions, we can confidently assume that relict hominids have no undercoat of any kind, just like any other primate. The actual length of the hair seems to vary from “short” (presumably a hair length of 1, 2 or 3 cm) to long. The hair has also been described as “thick” or “shaggy”, or sometimes even “sparse”. This could easily be explained by some form of moult, as described by the chief of the south-western part of the Xinjiang Uygyr Autonomous Region, who claimed that the “wild people” of the mountains went through an annual moult (“the hair falls in April”).

Almost every observer agrees that the hair on the head of Bigfoot has a different character from the hair on the body. It is usually very long, falling in cascades down over the forehead and the shoulders. It is usually tangled or shaggy. In some rare cases it has been described as more stiff, and often having a different colour from the rest. Females seem to have the longest hair.

The eyewitness accounts never mention any kind of beard or moustache; on the contrary, they often emphasise the lack of it. The colour of the skin on the face is often described as dark or reddish.

The head of *Homo troglodytes* is sometimes characterised as being very large and massive, and there are also descriptions of it having an elongated conical shape. Another set of data described the “wild man “ as having a very “thick neck”, the head being “bent forward”, and having a strong neck with powerful muscles, and the head being to some extent drawn into the shoulders. Some records even state that the head of the “wild man” is placed directly on the shoulders, that he has in fact no neck. This is very similar to the way the head is carried on the skeleton of Neanderthals. They have a very similar pose to the one described for Bigfoot.

As for the overall shape of the face of the relict hominids, it should be noted that observers generally tend to compare it with a human face. For example, Colonel Karapetyan described it as oval and chubby, and without any simian features, although it was extremely dark. Game warden Leontiev said: “In any case it was a person’s face, not an extended bestial muzzle.” Officer Kolpashnikova: “the face was like a very rough human face”. According to Sherpa Phurpa, it was a flat face of a beast with a lot of wrinkles, “very much like a monkey”. According to another Sherpa, the muzzle was “like a monkey” – “hairless and brown”.

From an anatomical point of view, the structure and shape of the forehead and brow area of the Bigfoot is very important. This area is also the most well known in fossil hominids, and the most thoroughly studied, which makes it possible to make a cautious comparison. A number of accounts describe the forehead as low and sloping. Some even compare it to the forehead of a dog. Athanasius Kircher notes the “wrinkled narrow forehead”. A Tibetan lama said that the “skull was very flat”. It should be noted though that these are in fact exceptions from the general rule. For example, a Tajik witness from the south-western outskirts of the Xinjiang, who was interviewed by Professor B.A. Fedorovich, said that the animal he had seen killed had an “almost straight forehead just like a man. It was not sloping in any way.”

Another important feature is the brow ridge. All higher primates, including fossil hominids and paleoanthropic humanoids, have prominent brow ridges. Often the witnesses did not describe the brow ridge as such, but noted the reverse side of the same phenomenon, i.e. deep-seated or deeply sunken eyes.

As for eye colour, there are some descriptive data, but not enough to be able to make any general conclusions, nor any comparison with fossil hominids.

It is also very important to describe the lower face of the Bigfoot as it seems to be unique. We must compare it to fossil forms to determine its place among the primates and answer a very important question: to what extent is the jaw prognathous, that is protruding forward, and are the teeth slanting?

One of the Caucasian informants has described the "wild man" as having a face like a man, but with lips extended forward so it looks like something between man and ape. V.A. Khakhlov has for instance mentioned witnesses saying that the "wild man" is a person, but the face is more like a snout. Geologist M.A. Stronin said: "the bear has a snout, but this creature did not have a forward pointing animal muzzle, it was much more rounded." Game warden V.K. Leontiev said something similar. "In any case it was a person, it did not have an elongated ferocious snout." There have also been some comparisons of the face of the Bigfoot. Sometimes it is described as having a "face like a monkey", but sometimes having a face like "both monkey and man", and even "did not see any monkey features in the face." In general it seems to be that the face of Bigfoot is flatter than a monkey's face, but not as flat as a human face.

It seems that prognathism is more pronounced in relict hominids than in humans, but less so than in monkeys. The lower jaw and the skull are more massive. The lower jaw protrudes more, and the teeth are slanted – according to the Kazakhs "like a horse". The teeth are also described as being very big, almost twice the size of human teeth. And finally the chin is retracted. In several cases it is also noted that the mouth of the "Bigfoot" is much broader than that of a human, and that the lips are thin or almost non-existent.

There is a certain amount of contradictory information regarding the size and shape of the nose of Bigfoot. Some witnesses describe a small sunken nose with large nostrils, whereas others talk about a rather large and almost human nose. There seems to be a similar large degree of diversity and shape of the external nose within Neanderthals. Finally, we can add that we do not have any descriptive information about the ears of Bigfoot. The very few descriptions are contradictory, from "pressed in" to "protruding".

In summary, there is a fair bit of variation in the data we have on the Bigfoot. This can be explained both objectively and subjectively. As we know, there is quite a lot of variation in the morphology of modern man as well as Neanderthals. This type of polymorphism probably gives some sort of biological advantage, so one would expect to find it in *Homo troglodytes* L. (Bigfoot) as well. Some populations will look more like modern humans than others. The extent to which details are being reported also depends on the subjective focus of the eyewitness and the mental state. Some features will be very clear to the observer, whereas others will be totally overlooked.

CHAPTER 12 • PRELIMINARY DESCRIPTION OF *HOMO TROGLODYTES* L. ("BIGFOOT")

(2) Biology
Although the anatomical structure of *Homo troglodytes* L. (Bigfoot) can be compared to fossil hominids, there is no way we can compare his lifestyle to the archaeological reconstructions of the lifestyle of the oldest and most ancient human ancestors. Each stage is usually characterised by specific forms of artificial stone tools found with their remains. Archaeologists clearly distinguish apemen with a stone "culture" (Pithecanthropus, Sinanthropus, Atlanthropus) from the stone "culture" of Neanderthals. Some even associate australopithecines with certain forms of bone tools, i.e. an archaeological culture. But as there are no specific finds associated with Bigfoot, it seems quite safe to say that he has no specific "culture" in any archaeological sense. As such there is no reason to consider investigating relict hominids by analogy with the lifestyle of the Neanderthal (Late Acheulean and Mousterian culture). We must study Bigfoot and his way of life solely as the lifestyle of an animal.

In Chapter 5, we gave an indication of the distribution of relict hominids – if not in the most recent time, then in a relatively recent historical past. Chapters 6-10 gave an overview of a variety of descriptive data from different geographical regions. Since the accuracy of the data is not constant, it may safely be assumed that some of the descriptions will be based on errors or folk legends. But the overwhelming majority consist of a considerable number of reliable observations done by well educated and cultured people. We also have the physical remains as described in the previous chapter.

Habitat
A typical habitat in a particular area, with a particular landscape, flora and fauna, is called a biotope. Some species are designed for a very particular type of area, whereas others can be found in very different habitats. These are called eurytopic species – such as foxes, golden eagles and ravens. Relict hominids belong to this group as well. The above descriptions of meetings and observations are of course evidence of this species being adapted to highly variable environmental conditions. The following main types of natural environment in the range of relict hominids are associated with meetings and observations:

a) We have to start with the high snowfields and glaciers. This is not because this landscape in actual fact plays any decisive role in the life of the relict hominids, but because the expression "snowman" has become so common, although it profoundly distorts the essence of the whole problem in the eyes of the general public. A great number of authors have written seriously about large apes being confined exclusively to the fields of eternal snow and ice, and this is of course biological nonsense. Dr. I. Sanderson has written to the author of this manuscript, that in his opinion the uncritical use of the epithet "snow" is the main reason it has been difficult to get this whole problem accepted as a serious scientific subject. And yet, from the hundreds of cases in the files I keep, not a single one even hints at "snow people" living in this ecological zone.

The epithet "snow" came into wide use because of the finds of tracks in the snowfields of the Himalayas. They were the first documented physical evidence of their existence, and they became available to the world of science as well as to the public. However, it soon became clear that the animal only occasionally crosses the alpine snowfields, and does not dwell in them. So the real question is, what makes these creatures appear on the snowfields and glaciers? And how much time do they actually spend in these areas?

They mostly appear here when they pass from one area without snow cover to another. There are two possible factors that might induce them to move around in the high-lying area with glacial moraines. Firstly there is the presence of salt in some of the lichens and mosses growing in this terrain. Secondly the ability of these same lichens and mosses to concentrate high levels of vitamins D and E during the winter. The same phenomenon can be seen in certain species of birds and rodents that move up to the snowline during reproduction, because the plants and the insects feeding on them in this area have a very high nutritional value.

The "snow people" may also move to higher areas to avoid dangers such as humans and possibly predators like wolves.

One also has to keep in mind that even in a severe winter some areas of higher ground can be free of snow, allowing rodents and hoofed mammals, and perhaps relict hominids, to dig for nutritious underground parts of plants – bulbs, rhizomes and tubers.

So, the "snow man" only appears in the snow temporarily in search of food, when going to another valley or when moving away from danger. There is no doubt that he is capable of spending quite some time in the area, and move around quite a lot. The tracks found in the snow, and on the rough and steep slopes of glaciers, should be taken as evidence of a high degree of adaptation to the snow and ice and the severity of the cold mountain climate.

It is interesting to speculate on what kind of biological tools make this possible. As the "snow man" belongs to the order of primates, it has no undercoat (underfur), and therefore not a particularly warm fur, although we can assume a significant seasonal increase in the amount of hair. The most important adaptation to moving in snow and ice is probably achieved by an increased development of the skin and subcutaneous tissue, especially a fat layer accumulated during the autumn as in many other animals.

It suggests that Bigfoot, like other higher animals, has the ability to starve for long periods. This gives him the opportunity to make transitions of large lifeless snow-covered areas. And yet it is very clear that the animal prefers to leave the snow as quickly as possible, as the tracks always move onto rocky or stony ground.

b) The most typical landscape for the Bigfoot is rocky. Though the "snow man" to a certain extent is adapted to movement through the snow, it is adapted for climbing to a much greater extent. A large number of eyewitness descriptions attest to the amazing dexterity with which the creature overcomes the most difficult sections of rocks and mountains, all completely inaccessible to humans, and to the ease with which the animal climbs up or down steep rock faces. Relict hominids are at least as well, or perhaps even better, adapted to the rocky terrain as the most common inhabitants of the rocky highlands like mountain goats and snow leopards. This area has a substantial food supply for rodents, ungulates and carnivores. At the same time it serves as a safe haven from various larger predators such as wolves.

c) The alpine stony grass field is also associated with many observations of relict hominids. It has a very abundant plant life, and thus food for a large number of animals in different seasons. There are abundant colonies of different mountain rodents – voles, pikas, marmots and so on – as well as grazing herds of wild mountain ungulates from deer to yaks.

d) We also have a certain number of data on relict hominids in subalpine shrub and forest zones. A series of observations from "bushes", "impenetrable forest", "thickets", "mountain forest slopes" suggests a very high population in these areas. Numerous references to bush and forest terrain, often associated with the valleys of mountain rivers, have been gathered in the Sayan Mountains, the Caucasus and in various other regions of Soviet Asia as well as outside the USSR.

e) A fairly large group of data describes relict hominids living in landscapes with sand and rocky desert areas. For the most part, this is mountain desert, that is located on pretty high altitudes, but this landscape also forces us to give up the idea of Bigfoot only living in mountains. In the deserts of Central and Middle Asia, the data points to Bigfoot not being confined to completely lifeless areas, but to those places where flocks of large mammals appear in certain seasons. Where there are herds of wild hoofed animals, there will also be bears and predators like leopards following them around. There will also be birds of prey, and scavengers on the corpses of dead animals. This is of course still a difficult area to live in, although Bigfoot is no more a permanent resident here than in the areas of permanent snow or ice. But it is another indication of the great adaptability of this animal.

f) The flat steppe landscape.

[Translator's note: there is no more text in this section – presumably the author never

got around to finishing it.]

g) Large reed beds are another kind a habitat for relict hominids. This refers above all to the vast lowland landscape surrounding Lop Nor Lake. These reed beds are home to a diverse wildlife. There are also indications of "wildmen" living in reed beds in the northern parts of Xinjiang, and a few other smaller areas.

h) There is also information about "wild men" from swamp areas such as the lower reaches of the Tarim River in Central Asia, and the foothills of the Talysh Mountains in the southern part of the Caucasus, as well as various other areas.

i) However, reed beds and swampy areas are in a sense only special cases of a much broader ecological aspect of the relict hominids, their association with water and ponds. Highly specialised mountain animals such as mountain goats have a very low need for water. They get their moisture mainly from eating plants, and can supplement their needs by eating snow. Thus they may not be seen near rivers and lakes for months. When it comes to relict hominids we have reason to believe that they, like any other alpine mammals, can get their moisture from plants, and so do not require frequent watering. On the other hand we have a number of data describing the animals wading in mountain rivers and streams. We have descriptions of female hominids bathing their young in mountain lakes and rivers, and quite a number of descriptions, especially from the Caucasus, of adult males and females swimming at night. We even have information indicating small populations living very close to mountain lakes in coastal thickets and in shallow water. (For example several families of *odes-Obi* living around the Iskanderkul and Payron Lakes near the Hissar Range.) We have every reason to believe that a relict hominid can be quite closely related to water, and that it, like many other species of primates, can swim well.

In summary we can find relict hominids in a huge variety of environmental conditions, from frozen glacial highlands to desert heat, from the extreme humidity of marshes to almost totally anhydrous sand dunes. The maximum height to find the Bigfoot is around five to six thousand metres above sea level, an area where humans cannot survive for long, but where the Bigfoot seems to be well adapted to live. The landscape where the relict hominids can be found is often extremely rugged and wooded and shrubby, although in other cases it is open and flat. This enormous variety of landscapes calls for an extremely varied locomotive ability, and a number of other physiological adaptations in the relict hominid. A relict hominid must have a high degree of adaptability to a range of very different environmental conditions. No other species of primate is capable of surviving in such a diversity of habitats except humans; but that is not so much a case of biological plasticity as a case of being able to alter the environment by building, and by constructing clothes and tools. Even under these circumstances relict hominids are in some sense vastly superior to humans, as they can live here without problems, whereas people even with modern equipment cannot survive for long.

We have already noted that despite the variety of habitats of the relict hominids they are to some extent spatially interrelated. One could argue for the existence of a single distribution area connected by corridors of migration or walks along mountain chains and uplands areas of Eurasia. This probably goes for America and Africa as well. Is there anything in common for these areas from an ecological point of view? Yes, they all have one thing in common – the basic biotope of hominids has a very small or non -existent human population.

This leads to a rather unusual ecological conclusion. The so-called anthropic factor – the relationship between a species and the human population – plays a major role for the relict hominids. This implies that the general biology of the species was formed relatively late in time, probably in close connection with the evolution and dispersal of humans across the globe. This of course pushed the relict hominids into the most inhospitable areas, and forced them to adapt to this environment by mobilising all their reserves of biological plasticity.

The evolution of this species is a very unusual one. It is a kind of contrast to the formation of new animals by domestication. In that case we see animals adapt to the conditions of life in close proximity to man. In the case of relict hominids, we see them adapt to the lack of human contact. They are in a sense the antithesis to a pet. The stories about people training or taming individual hominids may in no way be confused with the domestication of certain species.

V.A. Khakhlov talks about it like this: “According to the Kazakhs, the wild man can be found anywhere from the glaciers in the mountains to the sands in the desert and in reed beds near water. It is impossible to limit it to one particular habitat or situation. The only thing that is important is for the area to be deserted. It seems to me that we have all kinds of creatures in the mountains and the deserts. There are 'wild men' in the mountains, and there are the people of the reeds. They are separated by a belt of culture inhabited by modern man with his farming and agriculture. So there is a chance of meeting them in the mountains and in the reed beds. In both cases it can happen in summer, but in different months. In the alpine zone it usually takes place in June, when the continued presence of mosquitoes and midges in the valleys makes it impossible to stay there for long. It is only in this short period one can surprise a *ksy-gyik.* And this is simply because after the appearance of humans and livestock, the 'wild man' has nowhere to go. In the valley, when the reeds grow in spring, and the Kazakhs migrate from their wintering grounds, they usually do not see or hear anything relating to the *ksy-gyik.* In autumn there is sometimes talk about someone seeing the 'wild man' or his footprints, but after a short while,the 'wild man' leaves and disappears. 'Wild men' seem to avoid contact with people, and move away to other areas if nomads move into their normal habitat. When people live with their flocks in the valley, the *ksy-gyik* retreats to the mountains, or the other way to live in the reeds. Similar movements are known from many large animals.”

P.P. Smolin has mapped the distribution of information about meetings with relict hominids, and has made zoogeographical observations of great importance. Smolin has established that the meetings do not tend to take place in sparsely populated areas that are inconvenient and inaccessible to humans. He has shown that, if people historically settle along rivers, the meetings with relict hominids would gather around watersheds, with the greatest concentration around the joints of the watersheds *. Another, perhaps even more interesting, observation is the fact that the meeting with hominids also reflects a number of ethnic and ethnographic boundaries and divides. In some areas, the hominids seem to have settled in "no man's land" between the areas controlled by different ethnic groups. There is also information regarding what happened when these creatures were pressed towards the ultimate edge, the extreme northern boundaries of Eurasia. They did not retreat without a fight. There are various stories of fights, the creatures throwing rocks and shouting, etc. But of course in the end the humans won out.

The above mentioned observations partly explain the rarity of data at our disposal regarding observations of relict hominids. Most meetings are the result of completely unexpected encounters. Only very rarely does a hominid approach a tent, shelter or fire, or indeed a person, and usually only at night, perhaps because of what would be described as curiosity.

Earlier we talked about the landscape and ecological characteristics of the sites where relict hominids have been observed or where humans have met them, with consideration of whether the particular area was a permanent or temporary habitat of these creatures: the place where they would pass through on their migrations or whether it was a chance meeting with a random single individual. It would of course be desirable to be able to map cases of meetings and observations of single individuals (males and females), pairs, groups with or without young, or young by themselves. Unfortunately, the material is still not large enough to allow us to perform a genuine statistical and cartographic analysis. Our conclusions can only be indicative and preliminary in nature. In the existing series (a few hundred) of records of direct and indirect information on observations of relict hominids, at least 75% refer to observations of single individuals. The actual sex of the individuals is not always observed, but it is indisputable that the females are a minority. Of the remaining observations, the vast majority concern pairs, and almost always a male and a female. There are several observations of three individuals together – male, female and young (or male and two females). Groups of more than three individuals are very rare. In the stories told by local people, it is often said that these creatures live in "families".

In the chapters of this book giving an overview of descriptive material from different

* *Translators note:* It *is not entirely clear what the author means by the word "joint", but that is what the manuscript says.*

geographical areas, we very carefully recorded all references to observations of young animals. This has given us the opportunity to pinpoint what seems to be a few pockets of relict hominid breeding areas: in the south-west corner of Kashgaria; somewhere between Lop Nor lake, the southern part of the Gobi Desert, and the Nan Shan Range; Badakshan in Afghanistan; in the Hissar Range area, and perhaps in the Caucasus. These areas are not located at the highest altitudes, but they are quite deserted uninhabited places, isolated from humans. All the isolated breeding areas are clearly of a relict character. And yet it is in these areas that we have recorded the highest number of "family" observations.

All other observations can be divided into two major groups.

First of all we have the many observations of single individuals. The bulk of these are in all probability young individuals leaving the area where they grew up because of lack of food. This is a very common type of biological reaction for animals growing up. In these cases, the animals are probably highly mobile and can wander great distances away from their place of origin. It has already been suggested that the bulk of the individuals seen in the Himalayas are mostly younger animals and young adults with no offspring. We do not know yet what makes them appear in certain seasons on the slopes of the highest mountains in the world. But nothing seems to indicate that they climb the glacier areas. Up there we only meet mature and powerful individuals, who have left the breeding areas for the same reason – not enough food.

Secondly, we see pairs wandering in search of areas with favourable breeding conditions. These are usually seen in different geographical locations. They are probably victims of some kind of accident, forcing them to leave their normal breeding area.

Judging by recorded folk legends, some of which lead back to a distinct historical past, at one time "herds" or "tribes" of wild humans were much more common. The stories usually tell of the extermination of these "tribes" of wild people in the newly populated mountain regions after long and bitter fights. The "tribes" were forced out into more inaccessible areas. Modern observations of flocks are very rare exceptions, and are almost exclusively registered from the breeding areas.

In general relict hominids are extremely rare animals. It is of course very difficult to give even a preliminary assessment of the modern population of the species. However, there are sightings of a dozen or more individuals at some points, so we can probably assume a population of at least hundreds of individuals, but spread over an extremely large geographical area. Relict hominids are in other words extremely scattered, probably the most scattered living creatures on Earth. A main reason for this distribution is probably a lack of food in the modern world. According to Professor I.A. Efremov: "Every wild animal requires a certain area to feed itself and to ensure its existence for the entire duration of its life. The more barren and deserted the terrain, the larger the area needed, especially for big animals. A pair of swans need 2.5 km^2 of land and water for life and reproduction. A pair of ostriches need 24 km^2 of desert and steppe. A fox living in our temperate zone needs a lot less space than a desert fox. It is even more difficult for a large anthropoid ape-man. Eating fruit in a rain-forest, an ape will require a relatively small area. An anthropoid ape living in a mountain area will require an area many times greater to sustain life. If the ape lives in a really cold area close to the eternal snow of the mountain tops, just to keep its body temperature it will need a large amount of high-quality food, and this dramatically increases the area it needs." This argument has to some extent been used to show that the Bigfoot could not possibly exist. But this was when we only had observations from a small strip of land on the southern slopes of the Himalayas. Today we know these animals can be found over a much larger area.

In our opinion, relict hominids are to a large degree animals adapted to long migrations or wanderings. According to one informant from Mongolia, almas are like wild horses and kulans[56] a vagrant, that always moves around, and is usually never seen in the same place for long. This does not of course exclude the possibility of relict hominids staying in the same area for a longer period of time. We will study this aspect of their lives later, but first we must note how biologically fit to travel the creature is. This is demonstrated by a number of observations, such as the many descriptions of the long mammary glands of the females, and the fact that when walking or running they often throw them over their shoulders. Females of higher primates usually carry their young on their backs, and this enables the young to suck the mother's breasts while they are moving. In this way, females are able to keep up with the males without stopping to feed the young.

Moving in a vertical position enables the creature to carry various objects over long distances. Walking on the hind legs alone frees the front legs/arms for other tasks. We have in our material a number of references to the carrying of sticks, stones, large animal carcasses or parts of carcasses (legs, bones, hides). These objects can be carried using three methods depending on the objects: in the hand, under the arms, or on the neck while holding onto the objects with raised arms.

Generally speaking relict hominids are not especially well adapted to a settled life, although there are exceptions, although the amount of data cannot be considered long term. Informants often stress that the creature has no permanent dwelling. In a few cases there have been finds of nests lined with tangled branches, grass and so on. They have usually been compared with the one-day nests of the larger apes. Usually relict hominids use natural caves, pits and gullies for temporary shelters. In a few cases they can also use abandoned human constructions. Khakhlov writes: "The lair has been found under overhanging rocks, in hollows under thick bushes, and in niches in clay cliffs. There is almost always some form of lining consisting of large stems and grass. In reed areas, the creatures take shelter in pits or gullies covered with reeds. In a few cases, especially in Mongolia, herders claimed to have found dens, but without any kind of lining, only small animal bones. Other cases describe how a temporary dugout has been strengthened with stones, or a stone barrier has been erected to protect from the prevailing winds."

But we can still not come to a definitive conclusion regarding the total absence of relict hominids living a sedentary life. Professor Rinchen and Professor Dementiev gathered information about "hill almas" in south-western Mongolia. This may be an indication of a relatively large resident population of these creatures in the very recent past. The hills are overgrown with Haloxylon (saxaul) trees. There were very large holes dug out in them, and also vents for air. One of these hills, of which they have several sketches, was about 3.5 m, and the hole dug in it was about 1.6 x 1 m. When a local historian visited one of these artificial caves he found rabbit bones and human-like feces. According to old people of the area, these burrows, now mostly collapsed, used to be home to almas a few decades ago. Their local names were *nuhny almas*, which means cave almas.

Nutrition

Next we have to try to summarise the current data about the nutrition of the relict hominids.

Although only a few informants attempt to characterise the hominids as either "vegetarian" or "carnivore", the overwhelming majority of the descriptive data show us an omnivorous animal. It is not strictly vegetarian, as meat clearly plays an important and perhaps dominant role in the diet.

The fact that relict hominids are omnivorous is probably part of the reason they can be found in so many geographical areas and local conditions in different seasons. To some extent a relict hominid can be compared to the brown bear, which also demonstrates a variety of local food sources as well as individual skills and seasonal changes. Judging from the abundant descriptive material, a relict hominid is probably, like any other primate, primarily herbivorous and eats a very large selection of plants, although the collection of enough food for such a large animal must be very time-consuming. The only areas where this must be easier are some of the mountain slopes in places like Tien Shan, with an abundance of wild fruit and nut trees.

We have a number of descriptions of relict hominids eating "grass and leaves", "bamboo shoots", "some species of fern", rhubarbs and other species of plants, as well as certain species of mosses and lichens, various berries (e.g. mulberries), wild fruits and orchard fruits. Among the latter people have especially remarked on the creature's fondness for apples. A number of sightings of relict hominids, close to human settlements in the seasons of ripening fruits and vegetables, often describes them as eating plums, mulberries and various orchard fruits, melons in the melon fields, ears of corn and vegetables from fields and gardens. In the more desolate mountain areas, the main plant resource available to the relict hominids is not so much the visible parts of plants as the underground parts. They seem to be especially interested in the roots of rhododendrons and ginseng. Several observers have described how the relict hominids bend down to the ground, ripping and pulling at things, or “digging the ground and eating grass”, “pulling plants from the ground, cleaning the roots and eating them”.

The use of these underground plant parts - roots, tubers and bulbs - is particularly important in the most adverse winters, unless of course the ground is covered with a thick layer of snow. It has been suggested that relict hominids make stores for winter. This is probably a mistake. The animals probably plunder reserves made for the winter by prairie voles. Each of these larders can contain many kilograms of food. The hominids probably find them by following the tracks of the voles.

We have a certain amount of data regarding the eating of insects and worms, especially some form of mountain crickets. When searching for food, the hominids will turn over stones. The same goes for crabs and frogs, which according to some authors is a favourite food of the relict hominids.

We have only very limited information regarding the hominid's use of fish, birds and their eggs and chicks. The only information we have about catching fish in shallow waters and devouring them raw comes mainly from the area of Lake Lop Nor, and the lower reaches of the Tarim River. Incidents of catching wild birds are rare, and we do not know how a relict hominid may catch them. We do have more specific descriptions of hominids plundering birds nests and eating the eggs and young.

We have the most information regarding the relict hominid's use of a variety of rodents as food. It seems that they, along with the plant food, are an important food resource, at least in the modern era. Among the species eaten are relatively small rodents such as gerbils and rabbits, but the largest number of references describes them eating marmots and pikas. Marmots are fairly large (up to 57 cm in length and weighing up to 8 kg). Pikas are smaller (up to 30 cm in length) but very abundant, and apparently easily accessible to the hominids. Pikas live in vast colonies, are active year-round, and put in stocks for the winter in the form of hay which they hide in cracks between the stones. There are descriptions of hominids killing the pikas with stones. The marmots they either wait for near the entrance to their burrows, or dig out hibernating animals, as they are dormant for 6-8 months of the year. Sometimes they use sticks for the actual digging.

The animals do not seem to have an especially large need for drinking water, so they do not stay close to waterholes or visit them on a regular basis. When they have been seen drinking, they either suck up the water “like a horse”, or dip an arm in the water and lick it.

It would also be appropriate to say a few words about excreta and food residue from relict hominids. Quite a number of descriptions state that the excrement of “wild man” is very similar to humans. Due to the accumulation of excrement in the temporary lair of the relict hominids, it often emits a strong odour. Kazakhs claim they can not only find the lair by the smell, but also determine how old it is. Animal droppings are an important tool when studying the food of various species. According to V.A. Khakhlov, the droppings of relict hominids point to a predominance of plant foods, especially berries, but also contain remains of insects, hairs and bones. In the Himalayas Stonor and Tenzing noticed the remnants of small rodents, hairs and bones. In the same samples Stonor found large amounts of clay, corresponding to the stories told by Sherpas about Bigfoot having a habit of eating a particular kind of clay, probably for its mineral content.

In the book by Sanderson, there is a detailed review of laboratory tests of samples of relict hominid excrement. They used a widely known standard technique, including the identification of intestinal parasites.

First group: Excrement from *mi-te* from the Himalayas. Content: fur, bones, feathers (partridge?), remnants of various large insects. Shaped very much like human excrement.

Second group: Excrement from *those-lma* from the Himalayas. Content: remains of pikas, in addition to bones of frogs, plant residues and insect remains.

Third group: Very fresh droppings from California. Their huge size (up to 2 feet long) showed the difference from human excrement, but also excluded any North American mammal species. Research found three types of eggs from parasitic worms, including one species also found in humans.

In addition, we also have a few reports of food debris from relict hominids found on the ground: bones and scraps of carcasses, fur, tails or paws, and in one case the carcass of a marmot with the head bitten off.

Finally we have the strange habit of relict hominids not eating the insides of the animals they catch. No animal that I know of will ever do that. For instance, the entrails of pikas and marmots have been found around relict hominid footprints according to Sherpa reports. Stonor personally found the entrails of marmots on two occasions, as well as the entrails of pikas with a few tufts of hair. A thorough zoological analysis led him to conclude that these remnants could not be from the meals of birds of prey or any ground-living predator. They usually start by devouring the viscera. So, the only possible explanation is the one presented by the Sherpas. “It is unlikely that the Sherpas would come up with such detail in a fabricated story.” V.A. Khakhlov got the same type of information from Kazakhs regarding the *ksy-gyik*. “I heard twice, that they will gut the birds and animals before eating them, and throw the innards away. And it was told as an amazing phenomenon, a testimony to his intelligence and similarity with man.” One of these stories relates to the area of Manas, where a shepherd noticed two discarded set of innards from young turtle doves. The parent birds were sitting in a tree nearby, close to the deserted nest, and in the sandy soil he could clearly see footprints resembling humans, moving towards the coastal bushes and thick reed growth. Another story concerns an area where fresh entrails were found near a "wild man" burrow.

Such a large semi-carnivorous primate may not at all times be able to satisfy its need for meat by only catching small animals. A number of stories, especially from Kyrgyz people, say that they used to hunt wild yak, wild camel, wild horse and reindeer. According to G.K. Sinyavsky, the reduction in the number of "snow people" is connected to the disappearance of the wild yaks, that used to graze in great flocks on the alpine meadows. According to the Kyrgyz they were the main food of the *Aban* - hairy wild men. It is interesting to note that the south-western outskirts of the Xinjiang area, which we have identified as the main habitat and breeding area of the relict hominids, is also the area with the largest amount of remaining wild yaks, apart from a few areas in Tibet. Other pockets of relict hominids in Central Asia are found in the latest nature reserve for the wild camel.

Does this mean that we should regard this hominid as a hunter of large animals, a dangerous predator gradually dropping down to killing only small animals? On the one hand, there are a series of observations supporting such an assumption. According to some, when the snow is deep in the winter, it can supposedly hunt for mountain sheep or mountain goats and kill them with stones. It can also catch and eat sheep, calves, young yak and musk deer. One hunter allegedly saw a hominid fracturing the vertebra of a female mountain goat. It is also said to be able to kill adult yaks by breaking their spines. To this we can add a series of descriptions about how a "wild man" approaching can bring large domestic animals, camels, yaks, goats and so on to a state of mortal fear, perhaps a hereditary instinct reflecting the real danger.

On the other hand, it is still too little to construct an image of the relict hominids as ferocious hunters of large quadrupeds. If the described details are true, they only say that the relict hominid may sporadically be behind such attacks. The "wild man" was most likely never an aggressive predator; but it could not only dispose of the bodies of large animals that died of natural causes, it could also develop the skills and technique to finish off dying or weakened animals or deal with young or injured animals straggling behind the herd. Living in areas where early man was farming large ungulates, "wild men" would probably have enough food from the natural deaths of the livestock. We also have a story from a Chinese border guard about a Bigfoot carrying an entire cow carcass thrown out by people because it was unsuitable for human consumption.

So, one has to ask whether the seasonal migrations are in fact connected with the migrations of herds of ungulates. So far we do not have enough data to form any kind of conclusion. Many of the locations are very imprecise, and it is difficult to correlate the data with exact data on the seasonal migration of hoofed mammals in different districts.

A number of observations are rather strange inasmuch as they describe relict hominids in close contact with horses, often being taken for horse thieves in the dark. Horses have, as already noted, a fear of relict hominids, but there is no evidence that they have ever been hurt by hominids, except for the very common belief in many regions that "wild men" will climb on horses and ride them to exhaustion. Can there possibly be any form of evolutionary impetus behind the relict hominid's attraction to horses, obviously at first to wild horses and later transferred to domestic horses? There are a few stories of "wild men" knowing how to suck the milk of mares – often in rather elaborate ways reminiscent of trick horse riding as used by human riders. It is still premature to judge whether there is any real biological phenomenon at the core of these stories. It may be that the relict hominids once made large migrations following the wandering herds of wild horses. In any case, the biological ties between horses and relict hominids need further serious study.

Some very scant information has drawn researchers to the possible existence of an unknown biological connection between relict hominids and major land predators. There are for instance observations of Bigfoot and bears in close proximity to one another. There is a similar connection between Bigfoot and the snow leopard. In the case of the snow leopard it is also interesting that they live in the same biogeographical neighbourhood, and there is a close evolutionary and ecological connection. Both of these "snow" animals undoubtedly emerged in the Quaternary era, and adapted to the conditions of their current range more or less at the same time. The descriptive material relating to Bigfoot mentions that it is not afraid of the snow leopard. The latter on the other hand is afraid of the Bigfoot. How can they have mutual biological interests? For example, the presence of Bigfoot in an area could serve as a obstacle for settlement, and therefore to the destruction of the snow leopard. On the other hand, the antagonism of the snow leopard towards the common leopard, and most importantly towards the wolf, provides Bigfoot with the opportunity to live in areas normally out of bounds for them. Maybe the link lies in a different plane. Perhaps in times of abundance of hoofed mammals, the hunting snow leopards left a lot of animal food lying around, and the cunning primate recognised this as a food source.

All of these factors probably contributed to a greater or lesser extent to the evolution of the crepuscular lifestyle of the relict hominids.

Diurnal and Annual Cycles

The term "night person" (*Homo nocturnis*) is fully justified by the descriptive material and quantitative data we have at our disposal. We find a lot of generalising statements: "nocturnal", "appear only by day, if they have been frightened away from their hiding place", "sleep during the day, and leave their hiding place at dusk". The quantitative grouping of the individual observations is even more important. The few observations of sleeping relict hominids is always done in the afternoon, and the creature is always sleeping face down. In a large series of meetings with relict hominids that are awake, and where we have the time of day noted, less than 15% are during the day, more than 40% late evening and early morning, and the rest at night. In the words of V.A. Khakhlov: "I heard almost no stories about meeting the 'wild man' in the afternoon. The vast majority of meetings took place at dusk or at night. It seems that the *ksy-gyik,* like most other mammals, rests somewhere during the day, and at night goes out fishing or hunting." However we must remember that, apart from one subfamily of New World monkeys, no species of primate is nocturnal. All of them are day-living creatures like humans. This means that their senses and general physiology are deeply adapted to a diurnal way of life. So in this respect, the "night person" yeti would have to demonstrate an extreme degree of adaptability to change lifestyle like that. One of the more important aspects is of course a certain degree of functional adaptation of the visual organs to very low light, which is probably why the animals do not live in complete darkness, but prefer dusk, dawn and moonlit nights.

Apart from the daily life cycle of the relict hominids, it is also of great interest to get some understanding of the annual cycle of activity. As noted above, it is certain there is some degree of vertical migration with the seasons. But that is generally only in mountain areas. In other geographical areas such as forests or taiga conditions, there is no migration like that. For researchers, the most mysterious problem is the "disappearance" of relict hominids in winter from areas with sightings in spring, summer and autumn. In Russian folklore, there are stories of "evil spirits" that disappear with the first snow, and only come out again in the spring. However most scientists reject any idea of hibernation in relict hominids, mainly because it is not found in any family or species of known primate.

But it is worth looking into this question a little further. First we must distinguish between the phenomenon of hibernation and winter sleep, which is often intermittent and highly changeable. If there is a supply of food during the winter, even a minor one, there will be no hibernation, only winter sleep with periodic revivals. Secondly hibernation and/or winter sleep can be found to some extent in the most diverse group of mammals. It is not a specific characteristic of any particular family or group. Thirdly one must not hurry to the conclusion that, because hibernation is not found in any of the known primates, it is impossible for a single species to develop the required behavioural and biochemical mechanisms for it. The vast majority of primates live in warm areas, where there is no biological need for hibernation. However, there is one member of this group whose ancestors lived in cold conditions, and here we can actually see signs of just such a mechanism that is well understood and has been observed many times.

We are referring to the phenomenon of so-called lethargy in humans. Lethargic sleep has all the characteristics of winter sleep where complex life processes are reduced, so as to reduce the metabolic levels as well. If lethargic sleep in modern man can last a week or more, it is easy to imagine that for his evolutionary ancestors, where this was not a throwback or pathological disorder but a normal and biologically appropriate response, winter sleep could last much longer, and possibly only be interrupted for the occasional meal. The place of hibernation could be pits dug in the ground, possibly with loose earth and some form of leaf litter to work as a natural shelter from wind and snow, or they could sleep in caves. All this is of course only a hypothesis. *

Breeding

We will not dwell on the special characteristics of the breeding cycle of the relict hominid, because its basic elements have emerged in the previous presentation. But we need to add some details relating to the period when adult individuals have spent some time alone, wandering around whilst finally maturing, and have started to connect in pairs. For many species this stage of the biological cycle is very difficult and requires a lot of special adaptations. Butterfly males for instance find the females by their scent at a distance of several kilometres. Relict hominids have the added problem of being widely dispersed and spread over very large areas. This adds another threat to an already threatened species – the possible inability to find a mate.

The relict hominids apparently have developed two different adaptations to track and find their own kind, especially mates. The first is a special cry, and the second is a strong specific odour. Among the many different sounds observers have described under different circumstances, one in particular is audible at great distances in the mountains. It is usually heard just before nightfall. It is loud, high-pitched, very strong, usually prolonged, sometimes popping. There is no doubt that this cry has signalling value, as it seems to be adapted to the mountain echo. According to all observers, the cry of a relict hominid cannot be mistaken for any other animal.

Many animals have special glands that enable them to emit special smells and scents. This naturally begs the question of whether the very strong smell that has been reported several times in meetings with relict hominids has a similar function, although many just describe it as caused by “untidiness” or “sweating”. The sharp odour can be smelt at a considerable distance. The smell is so characteristic that there is a saying in Arabic: “stinks like a ghoul (*ghoul-biyaban*)”. It is quite possible that such a strong smell is some sort of signal, although it is so far impossible to state the physiological nature of the smell. If local people can find a Bigfoot simply by the smell, another individual would be able to do the same. However, the smell is not mentioned in all report of meetings with Bigfoot, so it may only be observable when individuals are searching for mates.

The use of screams and smells are not enough for every individual to find a mate, considering the current population of the relict hominids. This may explain stories

* *Translators note: There is in fact a biological difference between winter sleep and true hibernation, but the author uses the two terms completely at random.*

about how males have shown interest in girls and women, and females have shown interest in men. There are even stories, especially older ones, about abductions. This is probably a case of strong instincts being transferred to a cousin species, when the right partner is not available. This could also explain a series of folk-tales about fights between strong men and relict hominids. The same can probably be said about stories of relict hominids abducting human children. Maybe it is a manifestation of a deep-seated maternal care instinct that needs to be fulfilled.

Behaviour
Finally we will try to summarise information regarding the higher nervous activity of the relict hominids. They seem to have a considerably better developed sense of sight and hearing than humans. In the mountains they seem to be able to see and hear much better than humans, and this of course is one reason why it is so difficult to find them. They will see or hear people long before people can see or hear them. The sense of smell is somewhat reduced for primates in general, but a series of evidence points to the fact that the relict hominids still have a relatively highly developed sense of smell. Sherpas have emphasised that they are very sensitive to unfamiliar smells in an area, and easily scared. The smell of meat roasted on a fire, on the other hand, seems to be able to attract relict hominids from great distances, although it varies in different geographical areas.

It is difficult to say something about intelligence level in general, but many observations and stories refer to the highly sensitive, cautious and secretive nature of the relict hominids. They have been called "cunning", "clever", "no fool" and "much smarter than monkeys", "above all other creatures besides humans." But when compared to humans there is a certain degree of "stupidity", such as not being able to untie rope, some would run while still tied up, and some do not understand the danger of shots. When it comes to speed, agility and strength, relict hominids are far superior to humans. This could all be summarised in the words of a Kyrgyz: "He runs like the argali, sees like the golden eagle, hears like a leopard. There is no one stronger in the mountains."

There are a lot of available records about the different sounds and cries of the relict hominids. Most of it has been published at one time or another. Unfortunately it is very difficult to give this information an unbiased and objective treatment. It is notoriously difficult to describe sounds in writing, so it is also very difficult to compare various reports, although sometimes the differences can be explained by subjective perception or analogies. Yet if we compare all the available data it is possible to say that, apart from the very loud signalling sounds described above, relict hominids are capable of producing a great many other sounds: hisses, growls, miaowing, grunts, purring, bellowing and many others. The diversity of its vocalisations significantly exceeds that of chimpanzees and other large apes. On the other hand all the evidence describes these diverse sounds as "inarticulate". In other words, they cannot be called speech, not even rudimentary.

It is of great importance to study the behaviour of evolution of the relict hominids in relation to the many descriptions of them using stones, branches, poles and in one case a large femur of an animal, for various purposes. The almost complete liberation of the forelimbs because of the upright form of movement has made it possible for the animals to manipulate especially stones, available in great quantity in mountain areas. The many ways of using stones, as described by different observers, is quite impressive compared to the available scientific data concerning the respective capabilities of the larger apes. There is for example agreement to a large degree about the relict hominids being able to throw stones, often rather large ones, for a considerable distance. In ancient stories they are also supposed to have used stones in clashes with local tribes, especially the Kyrgyz. According to the older people in these areas many were killed by stones wielded by the relict hominids. We also have descriptions from caravans telling about humanoids throwing stones at them when passing through mountain passes. In the Pamir the population says that the wild hairy man has no fire and no guns, but knows how to throw stones and sticks. Residents of Hakimi village in the Hissar Range told G.K. Sinyavsky that “forty years ago wild hairy men from the mountains threw stones and boulders down the mountain and forced people to leave the place”. A similar thing happened in the Fergana district where wild men threw stones after people in one of the gorges.

In our material there is also evidence of the use of stones not only as a weapon against humans, but as a tool to kill animals. One hunter witnessed a wild man throwing stones at rabbits running through a gorge and killing one of them. Another set of observations describes how relict hominids use stones for different structures. They laid out stones and arranged them around a dug out pit-hole, and they made a little shelter from the wind with stones. Their stone-using skills are markedly different from other animals. A.M. Urazovsky writes that: “In the Abla Gorge in 1951, I saw three dug out marmot dens. They were not dug out by bears or wolves, as they will just throw stones anywhere. In this case the stones were neatly stacked to one side of the marmot hole.” When searching for frogs and large insects, they will sometimes move very heavy stones to one side. And they have been seen carrying two stones under their arms. Perhaps the most interesting information is a description of a wild man smashing a stone and using it to grind up grain.

What we see is of course not a “stone-based culture”, but it is still an example of a highly developed relationship with the material. Perhaps in the future we will know more of the relict hominid's way of using and manipulating stones, especially splitting stones, which would create sharp cutting edges useful for taking animal carcasses apart.

As for the emotive and neural stage of the relict hominid, we unfortunately only know something about reactions and behaviour when in contact with humans. First of all it is worth noticing a rather large number of observations of relict hominids making a sound resembling a laugh when coming in contact with humans. Or indeed making facial movements corresponding to the similar human ones, i.e. smiles.

Vercors has said that, if this is a laugh – a distinctive characteristic of man – then they are people on the same path as we are. This is an important realisation, as laughter is considered a "collision" (according to Pavlov) between two opposing stimuli or situations. Two reports also indicate that a relict hominid can become confused or frustrated. Twice one was seen (during a meeting with a car) to be frantically jumping on the spot ("dancing"). The emotion of anger or rage encompasses a whole range of expressions in relict hominids, including baring teeth, hissing and snarling. They have been described as having a similarity to a rabid dog. "Yeti began to run around the hut, trembling with rage, tearing up small bushes and twisting stones from the earth." "When *Hsiung-hara* is enraged, woe to the man who meets him at that moment. The *Hsiung-hara* will pull trees out of the ground with roots and all, and strike everything with them." As a contrast to this, a relict hominid caught in slavery was described as being in a state of profound nervous depression and apathy, whining pitifully or being passive and looking "dim and empty".

Two important aspects of the psyche of the relict hominids seem to manifest themselves in contacts with humans: imitation and curiosity. It is of course very interesting to compare these to the corresponding features of higher apes. When it comes to imitation, we actually only know this from folklore and legends; but the behaviour is described in such strong terms that it would be impossible to explain it in any other way than that it must once have taken place, have captured the imagination of people, and have therefore entered into the realm of folklore. This tendency to imitate is typically used by humans in the stories to trick and fight the relict hominids. There is much more extensive and realistic material concerning various manifestations of "curiosity" from relict hominids towards humans. In fact this phenomenon brings us to the question of the basic feature of the biology of the relict hominid – its "disengagement" from humans. They live in the most desolate and inaccessible places in relation to humans, where they adapt to the most severe conditions as long as they meet their basic requirement: minimum contact with humans. To some extent this is hampered by the extreme curiosity of the relict hominids towards humans, what Pavlov called the "What's that?" reflex or the investigatory reflex. A large series of data describes how relict hominids approach the tents and camps of humans and try to get close to unusual things, such as a fire – not from a desire to get warm of course, but from pure curiosity and partly a simple phototropism. We also have examples of relict hominids stealing objects from humans or breaking into empty buildings or even trying to dismantle the roofs of empty huts – all to satisfy their curiosity and answer the burning "What's that"? question.

It is easy to imagine that the same relentless curiosity towards humans will drive a relict hominid to use abandoned houses or sheds as lairs, and plunder insufficiently protected fields of maize, potatoes or melons, or indeed orchards. Even the occasional leftover food can satisfy this curiosity. In rarely used mills in the mountains, they can eat leftover grain and flour. This behaviour can sometimes be turned into an almost parasite-like connection with human settlements. But this only exists in areas where there is no chance for the relict hominids and humans to avoid each other. If given a chance, the relict hominids still prefer to stay as far away from people as possible.

The available data on the behaviour of relict hominids towards humans can be divided into three groups.

a) In the overwhelming majority of cases (at least 90%), the relict hominid will take flight when meeting people. As a rule the local population considers the hominids innocent, fearful and not physically dangerous creatures. People usually try to avoid meeting relict hominids, not because of any real danger, but because of the troubles and diseases they are sure such a meeting will bring. A belief that to some extent has basis in fact. There is a large number of data describing nerve shocks with severe after-effects and even death as a result of unexpected encounters with these animals. Muslim and Buddhist clergy have exploited these beliefs, so in most cases both parties are afraid and will flee from each other.

b) There is also a large amount of data about "curiosity" from one side or the other. There is a proportion of humans and relict hominids that will quietly watch the other party. An extreme expression of this behaviour can probably be found in the very few cases of alleged domestication of relict hominids. We do have descriptions from different geographical regions regarding the taming of relict hominids and the use of them for simple tasks.

c) In a very small number of cases, which have attracted great attention, we have reports on the aggressive behaviour of relict hominids towards people. The nature of this aggression is rather different. We see a series of similar stories about relict hominids attacking humans, possibly because they got too close to a breeding area, or because they started moving into a particular area. There are also a few descriptions of localised attacks, which supposedly have ended in the killing of a human and the subsequent devouring of the body. These cases of aggression can hardly be interpreted as proof of the existence of another form of bloodthirsty and warlike hominid, as opposed to the more peaceful and even cowardly form. In all likelihood, the behaviour of the same relict hominid can change according to time, place, and circumstances. The creature would naturally be aggressive when protecting family or offspring, or it could simply be irritated or angry for some other reason. However, as already mentioned, these forms of attacks are extremely rare.

Finally there is the possibility, or perhaps impossibility, of interbreeding between humans and relict hominids. There are a number of unconfirmed reports about the possibility of interbreeding and the appearance of the offspring. We note that these reports seems to point to the absolute dominance of most human traits. The descendants in the first generation are described as having the mind, speech and appearance of humans, whereas only a few features of relict hominids will manifest themselves. But in all cases these differences are no more significant than other minor differences between different groups of people.

CHAPTER 13 • DID THE NEANDERTHALS ACTUALLY BECOME EXTINCT?

Translator's note: It has been tough to decide what to do with this chapter – it is a discussion of hominid fossils, and the possibility of late survival of Neanderthals, based on the knowledge of fossils and human evolution at the time of the writing of this manuscript. Compared to our present knowledge of this, it seems outdated and obsolete, and I find it hard to see any value in this chapter. One might say that it has a certain value in a historical context, but I fail to see why. We know so much more today, so to keep this chapter would serve no other purpose than to point out they didn't know any better at the time. The author even states that this is not in fact a proper discussion, but more a question of him thinking aloud. And in the end he comes to the conclusion that even if a population of relict Neanderthals still exists, it is by no means proven that they have anything to do with the question of the Bigfoot. As long as nothing is proven, further discussion is meaningless. I have therefore decided that there is no reason to use time or space to publish this chapter.

CHAPTER 14 • MYTHS, LEGENDS AND FOLKLORE

In the previous three chapters we have looked at the available amount of descriptive and real data about Bigfoot and similar creatures, and tried to find a biological solution. Is it something unnatural or does it fit into a biological framework? Quantitative grouping of descriptive data is an unusual technique in zoology, but it has given a positive result. We have managed to formulate a rational preliminary morphological and biological description.

People who support a folkloric and mythological explanation for the yeti and equivalent beings typically base their ideas on the fact that many stories come over as legendary. But there are legends of all kinds of animals. There is for example a very large number of legends, fairy tales, beliefs and rituals related to the bear. What would happen, if anyone collected all these stories and ethnographic records of these beliefs, and tried to decide whether there is a real bear or not? In the absence of biological knowledge, they would of course have answered in the negative. The ethnographic record would for instance show that bears have many local names. This would force ethnographers to conclude that each name is associated with a specific "image". In some tales the bear can talk, in some it is the size of a mountain, it is incredibly frightening and vindictive if you do not appease it somehow. Meeting a bear can bring happiness, or unhappiness. The bear will carry a maiden to its lair – and so on. In short, the conclusion would be that the "bear", like the "yeti", is a myth, a fictional being, and that supporting people in their belief that the bear is real would mean supporting their superstition.

[Translator's note: The first four sentences of the next paragraph are particularly difficult. The language is very stiff and stilted, and I suspect it is in fact written as a political statement, or to appease any political official reading the manuscript. The sentence beginning "and makes it impossible to conclude..." I have translated rather faithfully. I think he is trying to be political here as well, and lamenting the fact that the section above does not make it possible to conclude with absolute certainty, that folklore is nothing but fiction.]

One could argue that the study of popular fantasies, myth and legends, although useful, can bring harm to science, as some ethnographers and folklorists have used it to give fiction a degree of validity. This is a dangerous trend in two respects. Firstly, it prevents the identification of those elements of truth in folklore that reflect the objective phenomena inherent in nature and history. And makes it impossible to conclude, that absolute pure fiction does exist within folklore. But handled in the proper way a folkloric study can reveal a lot of information and knowledge of the local fauna. But in far too many cases folklorists are not looking for information on the local fauna, but for things as unnatural and supernatural as possible. It would seem that ethnographers consider it boring to describe things that are not pure fiction. One cannot expect them to bring information from local hunters and shepherds about some new kind of animal, be it an unknown form of mouse or a great ape, to the attention of biologists. They only pick up their pen when they hear garbled and improbable information about the animals.

[Translator's note:There follows a section of short notes, where the author criticises a number of named folklorists and ethnographers for not taking the Bigfoot question seriously and, even when they do try to make an attempt at analysis, tend to pick out specific data that suits their own ideas – especially A.Z. Rosenfeld "whose work is characterised by inaccuracy, incompleteness and lack of attention to a number of details". The author does not in fact go into detail with his criticism, so I have decided to omit this section, as it leaves the reader with an impression of something that smacks more of a polite and scientific way of expressing personal grudges, and does not give the reader any useful information.]

From the material in the previous chapters we already know that the relict hominids go under a very large number of different names. Not only do different people give it different names, but individual persons often know it under different names. It would be of no interest to list all these names in this biological study. According to Sanderson there are hundreds of names for the creatures, but for convenience's sake he refers to them as ABSM for Abominable Snowman. But of course the number of names says nothing about the number of existing types of these creatures. It therefore seems somewhat inconsistent to attempt a philological and linguistic analysis of the names, as done by Sanderson and a young philologist named Rabbi Yonah ibn Aaron. This philological experiment does not add to the value of Sanderson's book, especially because more than half of the translations and interpretations of the included names are erroneous. Any analysis along these lines will only lead to confusion. The Bigfoot problem is not a linguistic problem. It is of no importance what this creature is called by uneducated people. They are themselves turning to the scientists for an explanation – what is it?

In most cases the name is simply descriptive and can be translated as “wild man”, “forest man”, “mountain-man”, “man-beast” or something similar, or the word “animal” can be substituted with “bear”, “wild goat”, “hyena” or similar. Then we get names meaning “man-bear” and so on. They always have the same similar kind of meaning – humanoid, but not a man, more like a wild animal or a spirit.

Linguistic examinations of the term “man-bear” have also been attempted, but they have been done very incompetently, as they usually attribute the various yeti attributes to bears. The most famous of these was done by the Indian scholar Swami Pranavananda. He tried to show that virtually all names did in fact mean “bear” or “brown bear”. The article by Swami Pranavananda was welcomed in the West among opponents of Bigfoot. The American anthropologists reproduced all the arguments of the Indian scientist in the journal *Science*. However, on closer inspection it turned out that Pranavananda's findings prove nothing in biological terms. According to zoogeographical data, the brown bear is not found in the areas where we have information about Bigfoot, and no bear can walk long distances “like a man”. Bernard Heuvelmans perfectly demonstrated the absurdity of the zoological representations in the Indian article. In the end it is only a linguistic result. But it does have a certain zoological interest that people in the Himalayas often refer to great apes as “man-bears”.

This is in fact confirmed by the considerably greater data relevant to the territory of the People’s Republic of China and several other geographical areas. The Englishman Bowden recently tried to examine the same “bear” concept on a new linguistic basis in *Man* magazine. Based on a number of corrections of a transcription done by Professor E. Vlček of texts with ancient Tibetan and Mongolian images of the “wild man”, Bowden came to the conclusion that the most important sign is the Chinese character meaning “brown bear”. He noticed that the Chinese term in itself does not prove anything, since it is in fact only an attempt at translation or interpretation of older names, such as the Tibetan *mi-kyo* and the Mongolian *Hsiung-guresu*. It seems that in Eastern China and in Tibet, as well as the Xinjiang Uygur autonomous Chinese region, “man-bear” is a very common name (*e-baths*, *zhen-hsiung* and others), as well as other forms of terminology closer to the bear, for the relict hominid. This is of course both philologically, historically and culturally interesting. From the sinologist

Prof. Oshanina we have obtained information about ancient Chinese "bear hunting" objects with depictions of creatures that do not look like a bear at all, but based on their erect posture more like an ape. Here the word is clearly at odds with the image. But words can have enormous power over the imagination and general thought processes. Rumours and miscomprehension of the ancient term could actually begin to ascribe observations not so much to relict hominids, but to beings intermediate between man and bear. This is in fact now widespread in some places in China. In other cases the spread of, for example, legends of bears and werewolves, can be turned into stories about humans and vice versa.

Primitive folklore is often populated with a large variety of divine and demonic beings. However, they would have been deprived of their specific features, disembodied and indistinguishable from each other. They would remain mere words and names, unless they were to be attributed to features borrowed from reality: people, animals or forces of nature. Perceptions of supernatural beings and phenomena have their own social and ideological roots, but many of the specific properties attributed to them are rooted in the real world. Sometimes they are mere words, names or an abstract title, but sometimes they take on a highly imaginative character. There is also a simultaneous counter-ideological process: supernatural properties are attributed to people, sometimes animals, and objects of inanimate nature.

When supernatural beings are given natural properties, they often borrow from different animals and humans at the same time, and end up as a bizarre mix, a clearly impossible combination of characteristics.

Is there the slightest reason to believe that the image of Bigfoot, as it appears before the data collector, is the product of such an artificial combination of features? Is it a combination of a human body shape with hair and no speech, such an improbable deviation from a real person? All these features can also be found in humans. It is worth noting in passing that some details in certain stories about creatures like the yeti seem to be purely legendary, although they may be explained as probable echoes of reality. For example, remember that in south-east China as well as Indochina, people who are at risk of meeting a humanoid creature in the mountains are told to wear a cuff or a tube of bamboo on the arm, so as to be able to pull out their arm and escape if the creature grabs them. This would seem to be nothing but improbable fiction. But recently I had a chance to read a description from a seasoned bear hunter named Mezintsev from the Khabarovsk territory. As the bears ran away at the sight of hunters with a gun, Mezintsev went hunting with only a dagger, and when the bear ran towards him and reared up on its hind legs, the hunter put one hand with a metal sleeve into its mouth, and killed it with the dagger in his other hand. Another story of seemingly pure fabrication comes from the Tajik (of Iranian origin). It tells of a man who enters into a contest with a *div*. The *div* shows him a very large louse, and the hero cheats it by showing it a frog, and claiming that it is a louse. But then one has to think about the observation of Colonel V.S. Karapetyan, who examined a wild man, and noted the abundance of lice – in particular on the face – and the exceptional size of the parasitic insects.

There is one thing that our critics might point out. Different people, in different parts of the range of the Asian "snow man", insist that the feet of the Bigfoot (under whatever name they know him) are the wrong way around, i.e. with the heels forward. This is a fairly minor thing for a supernatural being, but it is still a strange statement. Maybe people's imaginations are looking for things to make Bigfoot absolutely different from us? However, none of the persons actually claimed to have seen the twisted feet on an actual Bigfoot. Maybe the animal is more cunning than other animals and able to leave confusing tracks, or is simply capable of moving backwards?

It is undeniable that one relict hominid has given a wealth of material for many, many mythological images – most of them perhaps "evil spirits"; but many aspects can equally well be borrowed from the appearance or habits of wild animals and indeed human-like animals. One can anticipate a number of interesting discoveries in the study and comparison of those fantastic and half-fantastic images that populate the legends of so many people. But in the end this is a biological question, and the main phase of the research will be completed without any mythology added to the analysis.

We have already mentioned, for example, that stories of the *gesu tyrmak* can on the one hand be most incredible, and tell about demon-like bloodthirsty women with gold or copper nails, and on the other hand be perfectly sober: "Hunters killed three gesu tyrmak - a male, a female and a young female; their hands were the same as in humans. The hunters also found tracks that show these creatures are not afraid to cross water with a strong current". These differences can be very confusing, as the term "gesu tyrmak" strictly speaking means "demon". Similarly the word *almas* (*almasty*) was in the folklore canon supposed to designate only a certain kind of witch, but in our data it is used merely to indicate a wild hairy humanoid animal. The same goes for a number of other terms, all of them used to designate creatures from complete mystery to full reality.

To summarise some of the lessons of the failed folklore studies with which we began, we must emphasise that, contrary to the general belief in folklore, every "demonological" term does not have a strictly fixed meaning. Sometimes two or more terms may well replace each other. Sometimes the meaning of the same term is changeable and diffuse. Therefore, it would be impossible to study the overall question you are interested in just by the terms. If on the other hand you start with the image of relict hominid as described above, it is fairly easy to see how the creature inspired the different traditional expressions and images.

[Translator's note: the next few paragraphs are particularly difficult. I feel that some part is actually missing.]

According to A.Z. Rosenfeld there are three different elements we have to consider when studying these matters:

- the terms and names set forth in the various folk beliefs, legends and messages
- the images behind these terms
- the biological reality shown through these images

The number of terms and images used and known by the locals is enormous, even if we just look at the Pamir region. But are they all somehow connected? Rosenfeld tried to give each term a special value in an attempt to make a classification, but in the course of his studies he discovered that the various terms and names would gradually turn into other terms and names, so in the end he had to abandon the attempt to make a classification. Any type of connection could only be explained by some form of common biological background.

A truly mythological explanation for the Bigfoot would mean that all the first mountain people spread across Asia at exactly the same time, and that all of them developed very similar mythological stories about great apes, but with a lot of different names. But then one could also easily argue, that all the myths and images of ancient gods were based on the existence of a special group of people with hairless bodies and the ability to talk.

On closer inspection the mythological version turns out to be unprovable, contradictory and inconsistent with modern scientific methods. Provability and consistency is needed to accept that the yeti is actually existing, or existed in the recent past. Danserson writes that the study of myths, legends and folklore of these creatures is a rather boring affair. But it is in fact very important, because these myths, legends and folklore mean that almost everywhere, except in so far as we know Australia, Oceania and Antarctica, sub-humans, or perhaps even sub-hominids, could be found everywhere until the first *Homo sapiens*, who began to replace them or take over their territory.

Again we have three elements we need to consider. Firstly, a relict hominid will have made a large impression on popular beliefs in a variety of evil spirits. Secondly, stories about yetis coming from local people as well as some educated leaders are taken as folklore due to lack of information regarding the biological side of the problem. Thirdly, folk beliefs left a large impression on the population's attitude to the animal – turned it into a kind of totem or made it taboo. There is in fact considerable evidence pointing to a purely totemic relation to relict hominids. Here and there we find people who claim descent from Bigfoot and consider it their ancestor. Accordingly, a totemic animal is surrounded by mystery and omens. Seeing it is usually a bad sign, and observations of it often exaggerate its size or strength.

All this "primitive nonsense" can to a certain extent be observed in relation to a number of other wild animals. But it is immeasurably stronger in relation to relict hominids. As already stated above, the superstitions concerning the relict hominid have been supported and secured by Islam and Buddhism. They are ultimately to blame for the fact that he remained inaccessible to scientific investigation for so long. On the other hand, the relict hominids would probably not have survived to the present day without this superstition. It is necessary to know of these connections to myth, legends, folklore and beliefs for the sake of further practical field research. Researchers of relict hominids have to deal not only with the environment but also the humans. The attitude of local residents may be the greatest impediment and the greatest help in the search. To overcome this, one must be familiar with the prejudices and superstitions.

[Translator's note: There is a rather large but fragmented footnote to chapter 14 which basically says that, although folklore and religious and mythological interpretations do not have biological scientific value, it should be studied in much more detail to understand the way people interpret what they see, hear and experience.]

In his studies of the history of religion, E.M. Shtaerman has made a major contribution to the main problem. It is however somewhat paradoxical, because the author is just criticising his predecessor F. Böhmer for his one-sided interest in the religious beliefs of Roman slaves, without sufficient attention to the moral and other aspects of their social consciousness. But this broader approach helped Shtaerman to see in the beliefs of the lower classes of the Roman Empire something which is essential to the development of our science, the general theory of the origin and development of religion. The author establishes the tenacity of ancient communities and their local religious traditions. This gives them a link to a very distant past.

[Translator's note: Here follows a rather lengthy discussion of the development of religion during the Roman empire. The author mentions in passing once or twice that there is a certain resemblance between forest spirits and relict hominids, but does not pursue the subject further, just as he does not in fact state what major contribution Shtaerman actually did make. It is hard to see the relevance for the book, as is the case for Chapter 16. They appear to some extent as sketches for future books, which of course never came to be, so for the sake of clarity I have decided to omit them.]

CHAPTER 15 • THE WAY FORWARD: SEARCH METHODS AND TECHNIQUES

The general public seem to believe that there is no need to have any preliminary debate concerning the Bigfoot. We just have to catch one, and all will become clear. But, as we have stressed in the beginning of this book, a spark cannot produce an explosion if there is no barrel of gunpowder. It would be fruitless simply to hope for a single random capture or shooting of an animal. It will be of no value to science, because there would be no preliminary ideas or hypotheses about such a creature. Are we dealing with a pathological abnormality in a human individual (atavism, monstrosity, hypertrichosis, cretinism), or is it a simulation, perhaps a criminal hiding from the authorities, or even more absurdly some sort of "savagery" in people who have lived for a long time alone in the mountains; or are we even talking about "evil spirits"? Those who demand capture seem to be putting the cart before the horse. A capture will not be the beginning, but rather the result of a study. Where, when, who, how and why catch the creature? Without a reasonable answer to these questions, we will catch nothing.

Many people simply cannot imagine the difficulties one must overcome to catch creatures in the field. Difficulties which can only be overcome by very patient, lengthy and systematic fieldwork – even if we have found the right area. There are many reasons why it is extremely difficult to make deliberate observations of the yeti: extreme rarity; vagrancy; nocturnal lifestyle; instinct and skills, including sensitivity, rapid flight, adaptation to the landscape and a life that would be most difficult for humans; the absence of hunting prowess with these creatures in the local population that could help, and the generally unsympathetic attitude of the majority of the population in relation to such attempts at capture. There are also legal considerations regarding the presence of a dead or live humanoid and the attitude of local authorities who may not be familiar with our research. And finally, the question of whether it can actually be identified on the spot, and what category to put it in.

So what is the way forward for our research? Just to roam the infinite mountains and deserts of Asia, hoping to encounter a relict hominid? It would be like poking a spear randomly into water hoping to catch a pike. It is absolutely necessary to continue with the preliminary investigation and above all to expand the collection of descriptive material about meetings with and observations of the relict hominid.

We have to look into a clearer definition of a number of key factors: first and foremost, the natural habitat of relict hominids. This would lead to a better understanding of the fundamental biology of the species, and hence also any weaknesses in its behaviour where the human mind can triumph over its instincts and skills.

So, what is to be done? Especially in the field? Alas, impatient people will be disappointed to hear the answer. First of all, collect data from the local populations, especially the hunters, shepherds, herdsmen and people of other professions who spend time alone in the mountains and different deserted places. It can greatly increase our knowledge of whatever we are intending to test. This work will give an indication of the probability of observing the relict hominid in the particular region, and this will make it easier to choose the relevant area to search, and the specific conditions under which to conduct the search. If we are to rush into an area, ignoring the people living there and their knowledge, the amount of errors and wasted efforts will be much greater. The logical way forward seems to gain as much knowledge and cooperation from the local population as possible and proceed with that.

It is very likely that the Anglo-Saxon's arrogant attitude towards the Sherpas and their exaggerated ideas of their own abilities have to some extent prevented the solution of the question of the Bigfoot in the mountains of Nepal. In almost all descriptions of the Himalayan expeditions the Sherpas, who are excellent experts in their region and its fauna, are seen as almost completely passive observers when the "white sahib" searches for Bigfoot. They are neither materially or psychologically interested in the hunt being a success. Perhaps it is against their Buddhist faith? In other geographical areas, it must be possible to do more, and base the work on the experience of the local population.

Will it be necessary to offer a monetary award for the delivery of corpses of "wild men"? In most cases it would have no effect, because the people are not able to produce such a beast, as they would see the offer as unreasonable, seeing they believe that both the hunter and his children would die a painful death because of having performed an act of sacrilege. It could possibly have been effective in some parts of Asia. In the areas that we have identified as the probable focus of modern living and breeding relict hominids in the south-western edges of Kashgaria in China and south Tashkurgan, as based on the survey data above, the "wild man" is rarely a subject for hunting. But no matter how difficult it is to catch one, some people would probably try to take one for the high reward. However, we cannot recommend doing it like this. Getting hold of a single corpse, or perhaps a skeleton and skin, despite the importance of a comparative anatomical study, is not the main goal. This goal only seems a priority to those who see the problem as a kind of scientific wager with the sceptics. It is much more important to study the creature in its natural environment. Then, sooner or later, all the necessary material will come into the hands of researchers.

It is much more necessary to work with the locals. Perhaps to ask them to show the scientists the creature they are seeking. The expression "show" can mean a lot of things, from showing the researchers the places from where to observe the creature, perhaps take a photograph from a shelter – who knows. It may even be possible to tame one of the animals. This would of course demand changes in some very complex ideological and psychological processes in these people, if we want them to go beyond age-old traditions and superstitions. Any zoologist engaged in the study of relict hominids must also be a skilled teacher and promoter of a more scientifically based outlook. If successful, this is probably the most fruitful way for further discoveries. It is not easy, but not hopeless. The number of such people and the number of places in the vast mountains and deserts of Asia where cultural, ethnic and natural conditions would make it possible is small, but one must hope for success. If it were possible to find at least one place where residents still have some form of contact with the creatures as we have described earlier, it would be the key to a practical solution of the whole problem. From this it would be possible to expand with feeding experiments and further studies and contact.

There are of course other methods, but the question is, whether they would be of any use. One proposal is using a helicopter to find the "snowman". Others have tried to use a helicopter to find snow leopards or wolves or foxes, but had no luck. It is a hopeless task to see and photograph an animal that sleeps during the day somewhere in the bushes, rocks or caves, and only comes out at dusk or at night. Furthermore, it is a rare animal, scattered in small groups over thousands of square kilometres. And it is not even living on the plains, but in the mountains, where the helicopter would have to follow the terrain closely. At most you could expect to take photographs of tracks on various snow fields or in passes.

Equally impractical is the idea of stationing somebody with a large telephoto lens on a good vantage point. This point will inevitably be selected at random, and the chances of finding a rare animal in the field like this, for example a snow leopard, or indeed a much rarer humanoid, is extremely small. Besides, an observer would also need night vision equipment of some kind, as these creatures generally are active at dusk and dawn.

Using dogs to track the humanoids is not a viable idea either. First of all the dogs would not know what track or scent to follow, and as the humanoid has adapted to live in a terrain inaccessible to wolves and dogs, it would simply walk away from them and disappear.

It would of course be possible to wait for the creatures to show up at a particular location, but this would of course be based on local knowledge of a particular cave or a place of temporary residence for the relict hominids. The main drawback is the fact that this method requires a lot of labour with only a small chance of success.

Trying to lure the creatures with some form of bait is again a question of the right choice of location and the season. This should only be attempted in areas with a sufficient amount of oral testimony from the local population. For example, in late summer, when fruits ripen and fall to the ground, it is likely that a relict hominid will get closer to gardens and orchards. In the spring, after a winter of near starvation, the creatures are more likely to seek their food in the grassy rocky terrain with pikas and marmots.

From accounts collected in other areas we know of other things that might attract relict hominids. In the Qinling Shan mountains local people spread red cloth to attract the “wild man”, and there are a lot of stories about relict hominids being attracted to fire, and to the smell of roast meat, even from large distances. Maybe in the future we will also be able to record the cry of the relict hominids on tape, and play it back to attract other individuals. Any kind of bait should of course be surrounded by a band of loosened earth, making it possible to search for tracks that would confirm the bait is actually working.

So, imagine that with the help of local residents and bait we have finally made first contact with relict hominids. What comes next? The reader is probably already thinking of drugs, weapons or traps.

In fact, we can choose from a fairly large range of tools that could be applied now after the most difficult line has been crossed. It would be possible to resort to techniques of catching the relict hominid, that have long been used for catching monkeys. Namely, in a hole dug in the ground or a in container with a narrow neck you lay out fruit, so the animal puts his hand in there and grabs the fruit, cannot get it out again, and so can be captured. Or we can put out an open jar with barley beer or other alcoholic beverages. The animal will usually drink this and fall asleep on the ground.

It is of course also possible to replace these ancient techniques with more modern ones. It is possible to mix bait with any of the narcotic agents that are widely used nowadays for catching wolves and other large animals; one could use a crossbow with a bolt packed with a paralysing agent; it would be possible to build traps of some form, or finally use the most simple technique of them all: shoot the animal with a paralysing drug or a simple bullet.

A major problem is, that no one has thought about what to do after the capture or killing of a Bigfoot. There are major difficulties connected with getting a corpse from a deserted area to a scientific institution without it decomposing or deforming, especially the brain and spinal cord – invaluable for scientific study. There are also the internal organs and the soft tissue. And nobody seems to think about the fact that a specimen of a rare relict species if caught alive will almost certainly die before reaching civilization, as for instance snow leopards caught for zoos. Then of course one could examine the creature and release it into the wild again. It sounds scientific, but it would not meet the requirements of modern science as there would be no way of conducting proper tests and taking X-rays, doing physiological experiments or long -term observations. Obtaining “proof” can actually work against the true interests of science. This is not the way. Let us wait. We do not need to either catch or kill a relict hominid. It would be a far greater scientific victory to get a specimen used to taking food regularly. With the facilities of modern zoology, there would not even be the need for close contact. Colour films can be obtained by using telephoto lenses at a very great distance. There is also the possibility of infrared night filming and various recording tools. In this way the hominids could appear on a screen before the investigators in their natural environment showing their natural habits.

This should be sufficient to be able to start to study the higher nervous activity. Soviet physiologists of the Pavlovian school have developed and applied a special methodology of doing experiments with animals kept outside. You do not have to have a relict hominid in a laboratory – the laboratory itself can move closer to the hominid. First observations could recorded by filming and sound recording, TV and radio. And then one could study the impact of the behaviour by trying to develop new conditioned reflexes. This would lead to many scientific papers, and possibly to proper domestication of the animals. In this way the relict hominids could be protected and under constant scientific supervision. At some point one of these creatures would of course die, and then the anatomists could study the body, the skeleton, internal organs and tissue.

CHAPTER 16 • THE SECRETS OF ORIENTAL MEDICINE

Translator's note: This chapter is a very strange and incomplete one. It is, as the author states, "very short, but could easily be expanded to become a whole book." And "The study of this subject [folk medicine] may be the key to victory" [i.e. proving the existence of relict hominids]. The author's general thesis is that, as many kinds of animals have been used for their supposedly medicinal properties, it would be natural to assume that parts of relict hominids would have been used in the same way. That in itself is an interesting idea, which he expands to the suggestion that pieces of relict hominids may be found in medical stores, apothecaries and so on. But the author never really follows up on his theory. Unfortunately, most of the chapter is taken up with a discussion of a supposed wonder drug called "mummy" or "mumy" and similar variations, and its many miraculous effects, although there is not the slightest evidence of it being based on relict hominid parts. I have therefore decided to omit it, as even the few references are extremely obscure, and would be out of reach of most people. Anyone still curious can find a discussion of this drug and its source in Porshnev's book 'The Struggle for Troglodytes'.

CONCLUSION

Translator's note: The 4½ page conclusion in the manuscript is a another strange part of the book – not the least because the author starts out by stating that the only conclusion he can make regarding the "snowman" is that it is not a man and it has nothing to do with snow. The rest is not a conclusion in any way – it is very much a work in progress, and strangely enough written like a Platonic dialogue where he, "the author", discusses with an "opponent". Most of the questions however are sketch-like, as are the answers, and in general take the form of a philosophical discussion of what he is actually studying – physical remains or eyewitness descriptions, and whether these eyewitnesses are to be trusted. In the end, he arrives at what I would call a non-conclusion: the fact that there is no evidence ***against*** *the existence of the bipedal hairy relict hominid. On the other hand he realises that he has not actually proven anything, but by writing this book he hopes to have shown the way forward for further research. I have therefore decided to omit the whole section apart from this comment.*

NOTES

1 Bernard Heuvelmans, *On the Track of Unknown Animals* (London, 1958)

2 If this is the Malacca tapir, Cuvier himself credits it to his students Duvaucel and Diard.

3. Strictly, a colour variant of the cheetah and not a distinct species.

4. Sergei Obruchev was an explorer of north-east Asia and the author of numerous works in geology and geomorphology and of a number of travelogues.

5. According to a later book by Porshnev, *The Struggle for Troglodytes* (available as a PDF), the Study Commission lost official favour and the remaining volumes were never published.

6. Charles Stonor, *The Sherpa and the Snowman* (1955). Stonor was the scientific officer of the expedition; this is a very detailed analysis of not just the "Snowman" but the flora and fauna of the Himalayas and their people.

7. Russian zoologist who specialised in the study of mammoths and other Arctic paleofauna.

8. An assertion that Porshnev will repeat at the end of Chapter 4, though in neither place does he offer any justification.

9. W.C. Osman Hill, "Abominable snowmen. The present position", *Oryx,* v. 6, №2, August 1961, p.86

10. Odette Tchernine, *The Snowman and company* (London, 1961). See also her later book *The Yeti* (London, 1970) which draws heavily on Porshnev's work.

11. Ivan T. Sanderson, *Abominable Snowman: Legend come to Life. The Story of Sub-Humans on Five Continents from the Early Ice Age until Today* (Philadelphia and New York, 1961)

12. Rinchen was an ethnographer, linguist, scholar and a leading Mongolian intellectual. He collected almas accounts for more than 50 years.

13. P.K. Kozlov (1863–1935) was one of the most famous explorers of central Asia. He participated in the expeditions of N.M Przewalski, M.V. Pevtsov and V.I. Roborovsky. Then, from 1899 to 1901, he led an expedition in Mongolia and Tibet, and from 1907 to 1909, the expedition in Mongolia where he discovered the ancient city of Khara-Khoto.

14. Zhamtsarano (or Tseveen in Mongolia) was a Buryat Mongol and friend of Rinchen and Baradiyn, who studied Mongol history, law and literature and created the forerunner of the Mongolian Academy of Sciences. Caught in the Stalinist purges he died in prison in 1942, which is why any archives he left would not have been accessible to Porshnev.

15. Sergei Oldenburg (1863–1934) was a famous Russian orientalist whose works on Buddhism, ancient Indian literature and people of the far east are still authoritative. Continuing in his post after the Revolution, he became chairman of the Russian Geographical Society and later the permanent secretary of the Academy of Sciences of the USSR until 1929.

16. G.E. Grumm-Grzhimailo (1860–1930) was a geographer, orientalist, zoologist and explorer of central Asia, and author of many travel accounts. A glacier in Central Asia is named for him.

17. Sushkin and Menzbier were Russian ornithologists and members of the Academy of Sciences.

18. This episode actually appears further on in Chapter 4.

19. Harold W. Tilman, *Mount Everest 1938* (Cambridge, 1948), especially Appendix B.

20. Micheline Morin, *Everest from the first attempt to final victory* (Paris, 1953)

21. James B. Fraser, *Journal of a Tour through Part of the Snowy Range of the Himala Mountains, and to the Sources of the Rivers Jumna and Ganges* (London, 1820), especially pp. 234, 351, where however the word *bang* seems to refer to a cat-like creature or tiger [BP].

22. George F. White, *Views in India, Chiefly among the Himalaya Mountains* (1838).

23. Joseph D. Hooker, *Himalayan Journals, or notes of a naturalist* (1855).

24. William W. Rockhill, *The Land of the Lamas: Notes of a Journey Through China, Mongolia and Tibet* (New York, 1891) p.116 and 150.

25. Porshnev is either quoting from memory or using a bad translation. The last sentence actually says "This story of hairy savages I had often heard from Tibetans, while at Peking, and I was interested at hearing it again."

26. This date does not correspond with the dates given above. 1899 is the publication date.

27. Laurence A. Waddell, *Among The Himalayas* (London, 1899) p. 225. Rather than thinking no more about it, Waddell says that he interrogated many Tibetans on the subject,

but found no authentic case of a sighting; it was always 'something that somebody heard tell of'.

28. The report was actually published in The Times in 1921. The sighting is undated but is usually put in 1903. Again, Porshnev is quoting from memory or using a bad translation. Knight does not say the creature was terrible, only that it had 'terrific' muscles in arms and legs. He does not say the skin was covered in hair, only that there was hair on the head. However, there is some doubt that (William) Hugh Knight even existed and the report may be pure invention.

29. The *Statesman* is a Calcutta newspaper. Nicholas Roerich was the father of Y.N.Roerich and also travelled extensively in Asia.

30. In 1915 Henry Elwes read a letter to the Zoological Society from a correspondent stationed in Darjeeling describing a new species of ape or large monkey. Elwes did not claim to have seen the creature himself, and neither did his informant.

31 Lt.-Col. Charles K. Howard-Bury, *Mount Everest: the Reconnaissance* (New York, 1922), p.141. Howard-Bury, ignoring the opinions of the Sherpas, attributed the tracks to a loping grey wolf.

32 James R. Ullman, *Tiger of the Snows: Tenzing Norgay* (1955), p. 183.

33. William H. Murray, *The story of Everest, 1921-1952* (New York, 1953) p. 51.

34. Sir John Hunt, *The Ascent of Everest* (1953), p. 98–99.

35. Ralph W.B. Izzard, *The Abominable Snowman Adventure* (London, 1955)

36. According to Prof. A.N. Formozov, the presence of only four of the five toes in prints is not unusual [BP].

37. Norman Hardie, *In Highest Nepal. Our life among the Sherpas* (London, 1957)

38. Presumably a reference to the Basura Cave in Liguria. See Note 52.

39. A piece of disinformation that seems to have originated with an Italian provincial newspaper before being repeated in Tass, the Soviet news agency. Few countries in the world are less suited than Tibet for launching and recovering missiles: all known Chinese test facilities are in Inner Mongolia or at Lop Nor. The '600 people' were actually 20 personnel and 150 porters. The 'many missile experts' may refer to the radio-man Peter Mulgrew, who was a non-commissioned radio engineer in the New Zealand Navy, and Tom Nevison, a doctor working for NASA on space medicine. Certainly the Chinese suspected all such expeditions of espionage, and are said to have jammed the expedition's

radio communications!

40. Lawrence Swan was a Professor at San Francisco State University, had been born and raised in Darjeeling, and was an expert on Himalayan ecology.

41. It was actually the Khumjung monastery scalp that Hillary borrowed. The allegations of fraud against Hillary are of course completely baseless. Marlin Perkins, the team's zoologist, examined the Khumjung relic and realised that the material could not be from a scalp. Suspecting that it was actually from a serow, he had a mould made and shaped a section of serow hide over the mould. Both “scalps” were taken back for analysis and comparison. The original was eventually returned; the duplicate remains in the Explorers Club in New York.

42. Stanyukovich was presumably invoking the Principle of Competitive Exclusion, proposed by Russian microbiologist Georgy Gause in 1934 and applied to hominid evolution by Ernst Mayr in 1950, by which no two species can occupy the same ecological niche at the same time.

43. Rask, Raskeme, Rusk-Darya etc. refer to the Raskam river, a Kyrgyz name for the Yarkand or Zarafshan river. Part of its valley is also called Raskam. It rises in the Raskam Range which runs parallel with the Karakorum on the Pakistan border. The Kulan-Aryk is a tributary of the Yarkand.

44. Rybalko was a famous Soviet commander of armoured troops. He led 3rd Guards Tank Army in the assault on Berlin in 1945.

45. A Muslim anti-Bolshevik insurrection in Central Asia.

46. The 1928 expedition was joint German-Soviet. There were several Soviet expeditions to the Pamirs in the first half of the 30s. N.V.Krylenko was a very important Bolshevik politician who headed the Soviet legal system, where he propounded the view that political considerations rather than guilt or innocence should determine verdicts. A victim of his own judicial theories, he was shot in 1938. Presumably he led the Pamir expeditions in his capacity of head of the climbing association.

47. This sounds more like a poltergeist case.

48. The Vedas and the Zend-Avesta are not discussed in Chapter 2. The Avestas we have were compiled in the Sassanid period (not Achaemenid), 3rd-7th centuries AD. The dates of their author Zoroaster are uncertain to within four centuries, though a date of around 600 BC seems plausible. The 2nd Millennium BC date, let alone the 5-6000 year age, seems completely fanciful.

The Indian Vedas are discussed in Porshnev's book *The Struggle for Troglodytes*..

49. It is not entirely clear what victims or indeed what they are victims of. However, after the Crimean War the Russian government embarked on a program of ethnic cleansing against the Moslem tribes of the Caucasus. In 1864 the British consul C.H.Dickson toured Circassia and reported to the Foreign Office. In 1866 the consul William Palgrave made a similar tour of Abkhazia to collect information. Both reported the regions to be almost depopulated.

50. *Homo troglodytes* = 'Cave-dwelling man'. L = Linnaeus. Porshnev, or rather a colleague who critiqued the original manuscript, identified the relict hominoid with a species of this name originally described by Linnaeus. See *The Struggle for Troglodytes.*

51. It may seem surprising that Heuvelmans identified the skin with a species of goat associated with Sumatra, but at that point the two modern species of *Capricornis sumatraensis* and *C. thar* (the Himalayan serow) were not distinguished and were part of one species *C. sumatraensis thar.*

52 The fossilised footprints in the Basura cave at Toirano were carbon dated in 1972 and found to be only 12,000 years old. They belong to Homo sapiens of the Upper Paleolithic rather than Neanderthals.

53. Byrne himself says that he was allowed to remove a single finger from the hand in return for a donation of 10,000 rupees. The finger turned up in Osman Hill's possessions after his death and was transferred to the Hunterian Museum. Tested by geneticist Rob Ogden in 2011, it was identified as human by the DNA.

54 A reference to Chapter 3.

55. The Soviet anthropologist and archaeologist who devised the technique of reconstructing faces based only on the skulls, including those of Tamerlane and Ivan the Terrible.

56. A sub-species of the onager or Asian wild ass.

ENDNOTES

Page 12
The closely related Kodiak bear is slightly larger than the Kamchatka bear, and the polar bear larger still. 4 metres would be more likely length for a large white rhinoceros. Presumably the height of 2m 70cm for a standing mountain gorilla is a mistake for 1m 70cm.

Page 21
G.F.Debetz was a physical anthropologist who who studied the races of the Soviet Union and Afghanistan using such techniques as craniometry.

Page 27

The Bible: for example Isaiah 13:21 'But wild beasts of the desert shall lie there; and their houses shall be full of doleful creatures; and owls shall dwell there, and satyrs shall dance there'; or Isaiah 34:14 'The wild beasts of the desert shall also meet with the wild beasts of the island, and the satyr shall cry to his fellow; the screech owl also shall rest there, and find for herself a place of rest'. The Hebrew word *seir* or *saiyr* is translated as satyr in the King James Version, but as goat or wild goat in more modern versions; it appears in other books of the Old Testament where it can be applied to domestic goats and their young, to Esau (a 'hairy man'), and even, in Leviticus, to idols or demons.

Page 28
The *Indica* of Ctesias only exists in fragments quoted by later authors. Ctesias, who never visited India himself but based his account on traveller's tales, was notorious in antiquity for his extravagant stories. His inhabitants of India included one-legged men, men with tails, and people 18 feet tall. However, some of his information is genuine.

Page 28
Pompeii: this is mysterious. There appear to be no records or images of fighting apes or gorillas, certainly not on the amphitheatre at Pompeii.

Page 42
In *The Struggle for Troglodytes* Porshnev describes how he met the elderly Khakhlov in Moscow and interviewed him about his researches of half a century earlier.

Page 55
Although the story about British colonial troops is often repeated, there seems to be no solid basis for it. If it occurred at all it would have been in the early 1900s. Far from showing a lack of interest Sir Charles Bell, the British political officer for Sikkim, is said to have visited the site, inspected the body, and ordered it packed for transportation. However, there is no mention of it in his papers. Porshnev is often careless or even hostile towards Western sources.

Page 56
N.A.Tombazi, *Account of a Photographic Expedition to the Southern Glaciers of Kangchenjunga in the Sikkim Himalaya* (Bombay, 1925). The account is quoted at length in Heuvelmans, *On the Track of Unknown Animals.*

Page 56
The English pilot is Wing Commander E.B. Beauman, an outstanding pre-war mountaineer. He described his discovery of tracks during the 1931 expedition to climb a mountain called Kamet in a letter to *The Times* in July 1937, presumably responding to Kaulback's description mentioned below. Two days later a letter from Guy Dollman of the Natural History Museum suggested a large langur as a possible explanation.

Page 56
Kaulback: Articles in *The Times,* '20 Months in Tibet' and 'The Mountain Men' , 12 and 13 July 1937.

Page 57
Smythe: Article in *The Times,* 'Abominable Snowman. Pursuit in the Himalayas', 10 November 1937.

Page 57
Presumably the second paragraph refers to a letter in *The Times* by someone using the pseudonym 'Balu', who was almost certainly the climber Bill Tilman. The letter produced a correspondence between Smythe and Tilman (this time under his own name).

Page 119
Other versions of Topilsky's account (Porshnev's *The Struggle for Troglodytes;* Bayanov's *In the Footsteps of the Russian Snowman*) make it clear that the body was examined by the detachment's own doctor and not just by the Red Army soldiers.

Page 181
'...the first months of the Second World War'. Russians date the outbreak of the Second World War from the German invasion of Russia in June 1941, just as Americans date it from the Japanese attack on Pearl Harbour.

Page 181
In *The Struggle for Troglodytes* Porshnev describes how the Commission received an independent report of this incident from a friend of the commander of the patrol which had originally captured the creature. Dmitri Bayanov, in his book *In the Footsteps of the Russian Snowman* (published by CFZ Press 2015), offers a sequel to the story. He and his colleagues addressed an enquiry to the Minister of the Interior of Dagestan to provide information about the case. The Minister replied to the effect that, after 25 years of Socialism, there could be no 'wild men' in Russia; the creature must therefore have been a German spy and as such had been shot.

Page 184
Beginning in July 1942 the German Army Group A invaded the Caucasus, penetrating as far as the Caucasus Mountains and even sending a party to climb Mount Elbrus. They were forced to retreat in December after the German defeat at Stalingrad.

Page 213
Several caves in Israel and the West Bank are Neanderthal sites. However, none seem to match this description. Possibly Porshnev meant the Es-Skhul cave, which provided remains of ten individuals, or Qafzeh cave which had fifteen; originally regarded as Neanderthal they are all now seen as a very early form of Homo sapiens. Alternatively, excavation of the Shanidar cave discovered parts of ten Neanderthals, though that is in the Kurdistan province of northern Iraq.

Page 214
Despite his name Tschernezky was English, the son of Russian emigres, and a professor of zoology at Queen Mary College London. For his reconstruction of the yeti foot from Shipton's photograph see for example: "A Reconstruction of the Foot of the 'Abominable Snowman'", *Nature* 186, 496-497 (1960).

Index

The subject of this book is known by a multiplicity of local names. Consistent with Porshnev's view that nearly all of them refer to a single creature, they have been ignored in this index and replaced by the single neutral term 'relict hominid' (RH) based on Porshnev's original title for his book. Only the almas and the yeti, because of their historical importance in Eastern and Western studies, retain their own entries.

THE CENTRE FOR FORTEAN ZOOLOGY
www.cfz.org.uk

STILL ON THE TRACK OF UNKNOWN ANIMALS

The Centre for Fortean Zoology, or CFZ, is a non profit-making organisation founded in 1992 with the aim of being a clearing house for information, and coordinating research into mystery animals around the world.

We also study out of place animals, rare and aberrant animal behaviour, and Zooform Phenomena; little-understood "things" that appear to be animals, but which are in fact nothing of the sort, and not even alive (at least in the way we understand the term).

Not only are we the biggest organisation of our type in the world, but - or so we like to think - we are the best. We are certainly the only truly global cryptozoological research organisation, and we carry out our investigations using a strictly scientific set of guidelines. We are expanding all the time and looking to recruit new members to help us in our research into mysterious animals and strange creatures across the globe.

Why should you join us? Because, if you are genuinely interested in trying to solve the last great mysteries of Mother Nature, there is nobody better than us with whom to do it.

Members get a four-issue subscription to our journal *Animals & Men.* Each issue contains nearly 100 pages packed with news, articles, letters, research papers, field reports, and even a gossip column! The magazine is Royal Octavo in format with a full colour cover. You also have access to one of the world's largest collections of resource material dealing with cryptozoology and allied disciplines, and people from the CFZ membership regularly take part in fieldwork and expeditions around the world.

The CFZ is managed by a three-man board of trustees, with a non-profit making trust registered with HM Government Stamp Office. The board of trustees is supported by a Permanent Directorate of full and part-time staff, and advised by a Consultancy Board of specialists - many of whom are world-renowned experts in their particular field. We have regional representatives across the UK, the USA, and many other parts of the world, and are affiliated with other organisations whose aims and protocols mirror our own.

You'll find that the people at the CFZ are friendly and approachable. We have a thriving forum on the website which is the hub of an ever-growing electronic community. You will soon find your feet. Many members of the CFZ Permanent Directorate started off as ordinary members, and now work full-time chasing monsters around the world.

Write to us, e-mail us, or telephone us. The list of future projects on the website is not exhaustive. If you have a good idea for an investigation, please tell us. We may well be able to help.

We are always looking for volunteers to join us. If you see a project that interests you, do not hesitate to get in touch with us. Under certain circumstances we can help provide funding for your trip. If you look on the future projects section of the website, you can see some of the projects that we have pencilled in for the next few years.

In 2003 and 2004 we sent three-man expeditions to Sumatra looking for Orang-Pendek - a semi-legendary bipedal ape. The same three went to Mongolia in 2005. All three members started off merely subscribers to the CFZ magazine. Next time it could be you!

We have no magic sources of income. All our funds come from donations, membership fees, and sales of our publications and merchandise. We are always looking for corporate sponsorship, and other sources of revenue. If you have any ideas for fund-raising please let us know. However, unlike other cryptozoological organisations in the past, we do not live in an intellectual ivory tower. We are not afraid to get our hands dirty, and furthermore we are not one of those organisations where the membership have to raise money so that a privileged few can go on expensive foreign trips. Our research teams, both in the UK and abroad, consist of a mixture of experienced and inexperienced personnel. We are truly a community, and work on the premise that the benefits of CFZ membership are open to all.

Reports of our investigations are published on our website as soon as they are available. Preliminary reports are posted within days of the project finishing.

Each year we publish a 200 page yearbook containing research papers and expedition reports too long to be printed in the journal. We freely circulate our information to anybody who asks for it.

We have a thriving YouTube channel, CFZtv, which has well over two hundred self-made documentaries, lecture appearances, and episodes of our monthly webTV show. We have a daily online magazine, which has over a million hits each year.

Each year since 2000 we have held our annual convention - the Weird Weekend. It is three days of lectures, workshops, and excursions. But most importantly it is a chance for members of the CFZ to meet each other, and to talk with the members of the permanent directorate in a relaxed and informal setting and preferably with a pint of beer in one hand. Since 2006 - the Weird Weekend has been bigger and better and held on the third weekend in August in the idyllic rural location of Woolsery in North Devon.

Since relocating to North Devon in 2005 we have become ever more closely involved with other community organisations, and we hope that this trend will continue. We have also worked closely with Police Forces across the UK as consultants for animal mutilation cases, and we intend to forge closer links with the coastguard and other community services. We want to work closely with those who regularly travel into the Bristol Channel, so that if the recent trend of exotic animal visitors to our coastal waters continues, we can be out there as soon as possible.

Apart from having been the only Fortean Zoological organisation in the world to have consistently published material on all aspects of the subject for over a decade, we have achieved the following concrete results:

- Disproved the myth relating to the headless so-called sea-serpent carcass of Durgan beach in Cornwall 1975
- Disproved the story of the 1988 puma skull of Lustleigh Cleave
- Carried out the only in-depth research ever into the mythos of the Cornish

Owlman.

- Made the first records of a tropical species of lamprey
- Made the first records of a luminous cave gnat larva in Thailand
- Discovered a possible new species of British mammal - the beech marten
- In 1994-6 carried out the first archival fortean zoological survey of Hong Kong
- In the year 2000, CFZ theories were confirmed when a new species of lizard was added to the British List
- Identified the monster of Martin Mere in Lancashire as a giant wels catfish
- Expanded the known range of Armitage's skink in the Gambia by 80%
- Obtained photographic evidence of the remains of Europe's largest known pike
- Carried out the first ever in-depth study of the ninki-nanka
- Carried out the first attempt to breed Puerto Rican cave snails in captivity
- Were the first European explorers to visit the `lost valley` in Sumatra
- Published the first ever evidence for a new tribe of pygmies in Guyana
- Published the first evidence for a new species of caiman in Guyana
- Filmed unknown creatures

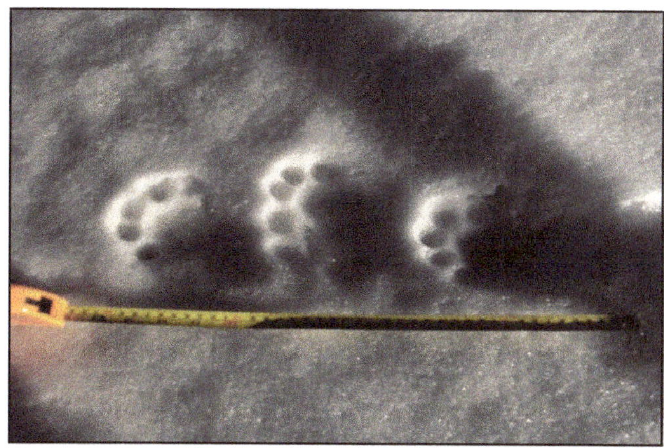

on a monster-haunted lake in Ireland for the first time

- Had a sighting of orang pendek in Sumatra in 2009
- Found leopard hair, subsequently identified by DNA analysis, from rural North Devon in 2010
- Brought back hairs which appear to be from an unknown primate in Sumatra
- Published some of the best evidence ever for the almasty in southern Russia

CFZ Expeditions and Investigations include:

- 1998 Puerto Rico, Florida, Mexico (Chupacabras)
- 1999 Nevada (Bigfoot)
- 2000 Thailand (Naga)
- 2002 Martin Mere (Giant catfish)
- 2002 Cleveland (Wallaby mutilation)
- 2003 Bolam Lake (BHM Reports)
- 2003 Sumatra (Orang Pendek)

- 2003 Texas (Bigfoot; giant snapping turtles)
- 2004 Sumatra (Orang Pendek; cigau, a sabre-toothed cat)
- 2004 Illinois (Black panthers; cicada swarm)
- 2004 Texas (Mystery blue dog)
- Loch Morar (Monster)
- 2004 Puerto Rico (Chupacabras; carnivorous cave snails)
- 2005 Belize (Affiliate expedition for hairy dwarfs)
- 2005 Loch Ness (Monster)
- 2005 Mongolia (Allghoi Khorkhoi aka Mongolian death worm)

- 2006 Gambia (Gambo - Gambian sea monster , Ninki Nanka and Armitage's skink
- 2006 Llangorse Lake (Giant pike, giant eels)
- 2006 Windermere (Giant eels)
- 2007 Coniston Water (Giant eels)
- 2007 Guyana (Giant anaconda, didi, water tiger)
- 2008 Russia (Almasty)
- 2009 Sumatra (Orang pendek)
- 2009 Republic of Ireland (Lake Monster)
- 2010 Texas (Blue Dogs)
- 2010 India (Mande Burung)
- 2011 Sumatra (Orang-pendek)
- 2012 Sumatra (Orang Pendek)
- 2014 Tasmania (Thylacine)
- 2015 Tasmania (Thylacine)
- 2016 Tasmania (Thylacine)
- 2017 Tasmania (Thylacine)

For details of current membership fees, current expeditions and investigations, and voluntary posts within the CFZ that need your help, please do not hesitate to contact us.

The Centre for Fortean Zoology,
Myrtle Cottage,
Woolfardisworthy,
Bideford, North Devon
EX39 5QR

Telephone 01237 431413
Fax+44 (0)7006-074-925
eMail info@cfz.org.uk

Websites:

www.cfz.org.uk
www.weirdweekend.org

THE WORLD'S WEIRDEST PUBLISHING COMPANY

NEW HORIZONS
Edited by Jon Downes

BIG CATS LOOSE IN BRITAIN

PREDATOR DEATHMATCH
NICK MOLLOY
WITH ILLUSTRATIONS BY ANTHONY WALLIS

Edited by Jonathan Downes and Richard Freeman

FOREWORD BY Dr. KARL SHUKER

A DAINTREE DIARY
CARL PORTMAN

THE COLLECTED POEMS
Dr Karl P.N. Shuker

STRANGELYSTRANGE
an anthology of writings by
ANDY ROBERTS

HOW TO START A PUBLISHING EMPIRE

Unlike most mainstream publishers, we have a non-commercial remit, and our mission statement claims that "we publish books because they deserve to be published, not because we think that we can make money out of them". Our motto is the Latin Tag *Pro bona causa facimus* (we do it for good reason), a slogan taken from a children's book *The Case of the Silver Egg* by the late Desmond Skirrow.

WIKIPEDIA: "The first book published was in 1988. *Take this Brother may it Serve you Well* was a guide to Beatles bootlegs by Jonathan Downes. It sold quite well, but was hampered by very poor production values, being photocopied, and held together by a plastic clip binder.

In 1988 A5 clip binders were hard to get hold of, so the publishers took A4 binders and cut them in half with a hacksaw. It now reaches surprisingly high prices second hand.

The production quality improved slightly over the years, and after 1999 all the books produced were ringbound with laminated colour covers. In 2004, however, they signed an agreement with Lightning Source, and all books are now produced perfect bound, with full colour covers."

Until 2010 all our books, the majority of which are/were on the subject of mystery animals and allied disciplines, were published by `CFZ Press`, the publishing arm of the Centre for Fortean Zoology (CFZ), and we urged our readers and followers to draw a discreet veil over the books that we published that were completely off topic to the CFZ.

However, in 2010 we decided that enough was enough and launched a second imprint, `Fortean Words` which aims to cover a wide range of non animal-related esoteric subjects. Other imprints will be launched as and when we feel like it, however the basic ethos of the company remains the same: Our job is to publish books and magazines that we feel are worth publishing, whether or not they are going to sell. Money is, after all - as my dear old Mama once told me - a rather vulgar subject, and she would be rolling in her grave if she thought that her eldest son was somehow in `trade`.

Luckily, so far our tastes have turned out not to be that rarified after all, and we have sold far more books than anyone ever thought that we would, so there is a moral in there somewhere…

Jon Downes,
Woolsery, North Devon
July 2010

CFZ Press is our flagship imprint, featuring a wide range of intelligently written and lavishly illustrated books on cryptozoology and the quirkier aspects of Natural History.

CFZ Classics is a new venture for us. There are many seminal works that are either unavailable today, or not available with the production values which we would like to see. So, following the old adage that if you want to get something done do it yourself, this is exactly what we have done.

Desiderius Erasmus Roterodamus (b. October 18th 1466, d. July 2nd 1536) said: "When I have a little money, I buy books; and if I have any left, I buy food and clothes," and we are much the same. Only, we are in the lucky position of being able to share our books with the wider world. CFZ Classics is a conduit through which we cannot just re-issue titles which we feel still have much to offer the cryptozoological and Fortean research communities of the 21st Century, but we are adding footnotes, supplementary essays, and other material where we deem it appropriate.

http://www.cfzpublishing.co.uk/

Fortean Words is a new venture for us. The F in CFZ stands for "Fortean", after the pioneering researcher into anomalous phenomena, Charles Fort. Our Fortean Words imprint covers a whole spectrum of arcane subjects from UFOs and the paranormal to folklore and urban legends. Our authors include such Fortean luminaries as Nick Redfern, Andy Roberts, and Paul Screeton. . New authors tackling new subjects will always be encouraged, and we hope that our books will continue to be as ground-breaking and popular as ever.

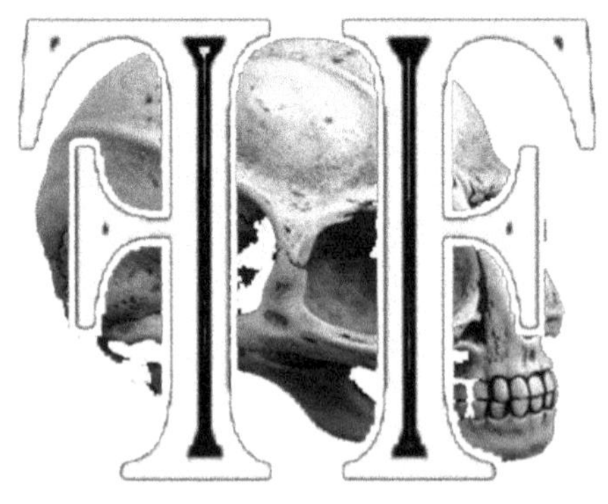

Just before Christmas 2011, we launched our third imprint, this time dedicated to - let's see if you guessed it from the title - fictional books with a Fortean or cryptozoological theme. We have published a few fictional books in the past, but now think that because of our rising reputation as publishers of quality Forteana, that a dedicated fiction imprint was the order of the day.

www.ingramcontent.com/pod-product-compliance
Lightning Source LLC
LaVergne TN
LVHW052344100826
845147LV00012B/748

* 9 7 8 1 9 0 9 4 8 8 6 4 9 *